ADVANCES IN EXPERIMENTAL MEDICINE AND BIOLOGY

Recent Volumes in this Series

Volume 625
DRUG TARGETS IN KINETOPLASTID PARASITES
Edited by Hemanta K. Majumder

Volume 626
GENOMIC IMPRINTING
Edited by John F. Wilkins

Volume 627
TRANSGENESIS AND THE MANAGEMENT OF VICTOR-BORNE DISEASE
Edited by Serap Aksoy

Volume 628
BRAIN DEVELOPMENT IN DROSOPHILA MELANOGASTER
Edited by Gerhard M. Technau

Volume 629
PROGRESS IN MOTOR CONTROL
Edited by Dagmar Sternad

Volume 630
INNOVATIVE ENDOCRINOLOGY OF CANCER
Edited by Lev M. Berstein and Richard J. Santen

Volume 631
BACTERIAL SIGNAL TRANSDUCTION
Edited by Ryutaro Utsumi

Volume 632
CURRENT TOPICS IN COMPLEMENT II
Edited by John D. Lambris

Volume 633
CROSSROADS BETWEEN INNATE AND ADAPTIVE IMMUNITY II
Edited by Stephen P. Schoenberger, Peter D. Katsikis, and Bali Pulendran

A Continuation Order Plan is available for this series. A continuation order will bring delivery of each new volume immediately upon publication. Volumes are billed only upon actual shipment. For further information please contact the publisher.

Stephen P. Schoenberger • Peter D. Katsikis
Bali Pulendran
Editors

Crossroads between Innate and Adaptive Immunity II

 Springer

Editors

Stephen P. Schoenberger
La Jolla Institute for Allergy
  and Immunology
Laboratory of Cellular Immunology
9420 Athena Circle, LA Jolla,
CA 92037, USA
E-mail: sps@liai.org

Peter D. Katsikis
Drexel University College of Medicine
Department of Microbiology and Immunology
2900 Queen Lane
Philadelphia, PA
USA
E-mail: peter.d.katsikis@drexel.edu

Bali Pulendran
Emory University
Emory Vaccine Center and Yerkes
National Primate Research Center
954 Gatewood Rd.
Atlanta, GA 30329
USA
E-mail: bpulend@rmy.emory.edu

ISBN: 978-0-387-79310-8   e-ISBN: 978-0-387-79311-5
DOI: 10.1007/978-0-387-79311-5

Library of Congress Control Number: 2008931353

Preface

This compilation contains minireviews derived from work presented at the Aegean Conference: "Second Crossroads between Innate and Adaptive Immunity," which took place from June 17 to 22, 2007 at the Aldemar Knossos Royal Village on the island of Rhodes, Greece. The goal of this conference series is to understand how the initial innate immune response that recognizes ancient patterns of infection can inform and influence the subsequent adaptive response that learns and focuses on a subset of idiotypic antigens. Sessions included those dedicated to innate sensing mechanisms; the immunobiology of T, B, NK, and dendritic cells; and the molecular and cellular control of immune responses; and featured presentations by leading scientists in each of these areas. In what is becoming a tradition of the "Crossroads" meeting, the participants were treated to an especially dynamic and interesting exchange of new data and compelling ideas, and the fruitful discussions that took place will no doubt set the stage for future experimentation in this rapidly evolving area of immunology, thereby providing the subject matter for the next iteration of this unique conference.

San Diego, CA	Stephen P. Schoenberger
Atlanta, GA	Bali Pulendran
Philadelphia, PA	Peter Katsikis

Acknowledgments

We are grateful to all the participants of this conference whose enthusiasm and willingness to share new data and discuss new ideas truly represent the lifeblood of the "Crossroads" experience. Those who have contributed chapters to this book deserve an extra measure of our gratitude, as we are keenly aware of the dedication they have shown to this field through their extraordinary efforts in this regard. We are indebted to our colleagues at Springer for giving us the support and opportunity to publish these proceedings in their Advances in Experimental Medicine and Biology book series, with special thanks to Melanie Wilichinsky for her sincerely appreciated efforts in assembling and formatting the chapters. We are once again deeply grateful to the Aegean Conference organization for the superb job they do in making this meeting possible, and especially to Dimitrios Lambris for his outstanding organization, logistical, and personal skills that make this meeting such an effortless pleasure for the three of us. Finally, we wish to extend our heartfelt thanks to the various entities that made this conference possible through their generous support, including the Aegean Conferences, Abcam, The Institute of Infectious Disease and Molecular Medicine at Drexel University College of Medicine, Drexel University, eBioscience, Kirin-Pharma USA (formerly Gemini Science, Inc), Genentech, Inc., Novartis, the *Journal of Rheumatology*, and Wyeth Pharmaceuticals. We look forward to seeing all of you at the next meeting in 2009.

Contents

Contributors

Kazumichi Abe
Department of Medicine, University of California at San Diego, La Jolla,
CA, USA

Sophie Agaugué
Istituto Giannina Gaslini, L.go G. Gaslini 5, 16147 Genova, Italy

Kory L. Alderson
Department of Microbiology and Immunology, University of Nevada Reno
School of Medicine, Reno, NV 89557, USA

Christine M. Bucks
Department of Microbiology and Immunology, Drexel University College of
Medicine, 2900 Queen Lane, Philadelphia, PA 19129, USA

Craig P. Chappell
Department of Immunology, University of Washington, Box 357330, Seattle,
WA 98115, USA

Mariella Della Chiesa
Dipartimento di Medicina Sperimentale, Sezione di Istologia, Via G.B. Marsano
10, 16132 Genova, Italy

Joseph Dauner
Department of Microbiology and Immunology, Emory University, Atlanta,
GA 30329, USA

Jose Gonzalez-Navajas
Department of Medicine, University of California at San Diego, La Jolla,
CA, USA

John T. Harty
Department of Microbiology, Interdisciplinary Graduate Program in Immunology,
University of Iowa, 51 Newton Road, 3-512 Bowen Science Building, Iowa City,
IA 52242, USA

Daniel J. Hodson
Laboratory of Lymphocyte Signalling and Development, Babraham Institute,
Cambridge, UK

Joshy Jacob
Department of Microbiology and Immunology, Emory University, Atlanta,
GA 30329, USA

Kyoko Katakura
Department of Medicine, University of California at San Diego, La Jolla, CA, USA

Peter D. Katsikis
Department of Microbiology and Immunology, Drexel University College of
Medicine, 2900 Queen Lane, Philadelphia, PA 19129, USA

Yuki Kinjo
La Jolla Institute for Allergy and Immunology, 9420 Athena Circle, La Jolla,
CA 92037, USA

Mitchell Kronenberg
La Jolla Institute for Allergy and Immunology, 9420 Athena Circle, La Jolla,
CA 92037, USA

Je-Wook Lee
CGK Co. Ltd, Daejeon Bioventure Town, Daejeon 305-811, South Korea

Jongdae Lee
Division of Rheumatology, Allergy and Immunology, University of California
at San Diego, La Jolla, CA, USA

Kelvin P. Lee
Department of Immunology, Roswell Park Cancer Institute, Buffalo, NY 14263,
USA

Emanuela Marcenaro
Dipartimento di Medicina Sperimentale, Sezione di Istologia, Via G.B. Marsano
10, 16132 Genova, Italy

Alessandro Moretta
Dipartimento di Medicina Sperimentale, Sezione di Istologia, Via G.B. Marsano
10, 16132 Genova, Italy
Centro di Eccellenza per le Ricerche Biomediche, Università di Genova, V.le
Benedetto XV, 16132 Genova, Italy

William J. Murphy
Department of Microbiology and Immunology, University of Nevada Reno
School of Medicine, Reno, NV 89557, USA

Jayakumar R. Nair
Department of Immunology, Roswell Park Cancer Institute, Buffalo, NY 14263,
USA

Silvia Pesce
Dipartimento di Medicina Sperimentale, Sezione di Istologia, Via G.B. Marsano
10, 16132 Genova, Italy

Bali Pulendran
Emory Vaccine Center, Atlanta, GA, USA

Eyal Raz
Division of Rheumatology, Allergy and Immunology, University of California
at San Diego, La Jolla, CA, USA

Erika D. Reynoso
Division of Medical Sciences, Harvard Medical School and Department of Cancer
Immunology and AIDS, Dana-Farber Cancer Institute, Boston, MA 02115, USA

Cheryl Rozanski
Department of Immunology, Roswell Park Cancer Institute, Buffalo, NY 14263,
USA

Stephen P. Schoenberger
La Jolla Institute for Allergy and Immunology, Laboratory of Cellular
Immunology, 9420 Athena Circle, LA Jolla, CA 92037, USA

Shannon J. Turley
Department of Pathology, Harvard Medical School and Department of Cancer
Immunology and AIDS, Dana-Farber Cancer Institute, Boston, MA 02115, USA

Martin Turner
Laboratory of Lymphocyte Signalling and Development, Babraham Institute,
Cambridge, UK

Thomas Wirth
Department of Microbiology, University of Iowa, 51 Newton Road, 3-512 Bowen
Science Building, Iowa City, IA 52242, USA

The Protective Effects of Type-1 Interferon in Models of Intestinal Inflammation

Jongdae Lee (✉), Kazumichi Abe, Kyoko Katakura, Jose Gonzalez-Navajas, and Eyal Raz

1 Introduction

Inflammatory bowel disease (IBD) including both Crohn's disease and ulcerative colitis is characterized by episodes of intestinal inflammation that can reoccur throughout life. Clinical and experimental evidence suggests that the etiology of IBD is multifactorial and includes susceptibility genes related to innate immunity (e.g., Nod2, IL-23R) as well as environmental factors,[4–6,10,13,14] such as exposure to intestinal microflora and/or their products. It is believed that the interaction of these factors with the immune system leads to dysregulated mucosal immunity and chronic intestinal inflammation. Different models of experimental colitis have been instrumental in defining the fundamental molecular mechanisms and cellular interplay that lead to colonic inflammation in mice as well as humans.

The acute colitis that is induced in mice by a single oral administration of dextran sulfate sodium (DSS) is the result of a T cell independent inflammatory reaction. This model is characterized by colonic epithelial-cell death (e.g., ulcers), mucosal edema, and, consequently, the accumulation of neutrophils that are necessary to limit bacterial translocation to the adjacent tissue. Subsequently, phagocytic monocytes and macrophages accumulate to remove dead cells (e.g., neutrophils) and tissue debris, and help to restore the physiologic function of the inflamed mucosa. Our previous analyses have identified type-1 interferon (IFN1), produced by toll-like receptor (TLR)-activated cells, facilitates the resolution of DSS-induced colitis. Here, we summarize our studies that have identified this novel function of IFN1 in the maintenance of colonic homeostasis.

Jongdae Lee
Division of Rheumatology Allergy and Immunology, University of California, San Diego, La Jolla, CA, USA

S.P. Schoenberger et al. (eds.) *Crossroads between Innate and Adaptive Immunity II*, doi: 10.1007/978-0-387-79311-5_1, © Springer Science + Business Media, LLC 2009

2   Toll-Like Receptors

TLRs are the key sensors of microbial invasion in mammals.[2],[3] They activate a defense program vital for host survival. Indeed, most of the phenomena associated with infection are traceable to TLRs. In mice, there are 12 members of the TLR family (TLRs 1–9, 11, 12, and 13), whereas there are 10 TLRs (1–10) in humans. Each of these TLRs senses a particular subset of signature molecules of microbial origin. While some of the TLRs reside at the cell surface (e.g., TLRs 1, 2, 4, and 6), other TLRs (e.g., 3, 7, 8, and 9) are located within the endoplasmic reticulum. These latter TLRs interact with microbial nucleic acids; TLR3 is triggered by dsRNA, TLR7, and TLR8 by ssRNA, and TLR9 by unmethylated CpG DNA sequences. Four adapters have been identified as key transducers of TLR signals. These adapters operate in functional pairs. MyD88 can either homodimerize to signal or does it in conjunction with the second adapter Tirap. The third adapter, Trif, signals from TLR3 as a homodimer. Trif and Tram, the fourth adapter, signal from TLR4, which also utilizes the other two adapters, MyD88 and Tirap. These adapters lead to activation of protein kinases that then activate the NF-κB, MAPK, and IRF families of transcription factors. Of all the TLRs expressed in mice and humans, only four (TLRs 3, 4, 7, and 9) induce the production IFN1.

In addition to traditional innate immune cells (e.g., dendritic cells or macrophages), intestinal epithelial cells (IECs) also express several TLRs including TLR2, TLR3, TLR5, and TLR9. Recent reports from our and other laboratories have demonstrated that activation of certain TLRs on colonic IECs can limit inflammatory responses to the myriad number of other TLR ligands in the intestinal lumen.[9]

Although intestinal microbiota, under certain conditions, provoke intestinal inflammation, they also support colonic homeostasis, mainly via TLR-induced repair genes. We and others have recently observed that the administration of certain TLR ligands, especially synthetic TLR9 ligand, effectively reduces the severity of colonic inflammation in DSS and other models of experimental colitis.[11],[12] When used in a preventive mode, administration of TLR9 or TLR3 ligands ameliorates DSS-induced colitis mainly via the induction of IFN1. In this review we describe our observations using various models of experimental colitis that led us to assume an important and novel role for IFN1 in intestinal homeostasis.

3   Probiotics and Experimental Colitis

Probiotics are live commensal microorganisms of the intestinal tract that confer health benefits to the host. Currently, probiotic therapy is advocated mainly for its immunomodulatory and anti-inflammatory activities at mucosal sites. The rationale for using probiotics in IBD is based on evidence implicating enteric bacteria in the pathogenesis of various models of murine colitis and IBD in humans. Indeed, probiotic therapy has been effective for the attenuation of experimental colitis, prevention of pouchitis, and maintenance of remission of pouchitis and ulcerative colitis.

Despite these beneficial effects, the exact mechanisms and the molecular pathways by which probiotics ameliorate experimental colitis and IBD are largely unknown. Improvement of intestinal barrier function, diverse and beneficial metabolic activities, competitive exclusion of indigenous intestinal microflora, and activation of the mucosal immune system have all been implicated in mediating the therapeutic effects of probiotics. In a recent study, we provided biochemical, immunologic, and genetic evidence that implicated TLR signaling, especially TLR9, in mediating the protective effect of probiotics (VSL-3) on various models of experimental colitis.[12] The administration of γ-irradiated (i.e., nonviable) probiotics effectively ameliorated experimental colitis, as did the administration of viable probiotics. Since the irradiated probiotics were unable to grow in culture, it is unlikely that either their metabolites or their competitive inhibition with indigenous microflora were responsible for the protective effects on the colonic mucosa. Therefore, we reasoned that the observed anti-inflammatory activities could be due to the activation of innate immunity (e.g., via TLR) by structural probiotic components (e.g., via TLR ligands).[11] To further verify the role of TLR signaling in the probiotic-induced amelioration of experimental colitis, TLR2$^{-/-}$, TLR4$^{-/-}$, TLR9$^{-/-}$, or MyD88$^{-/-}$ mice were treated with DSS and irradiated probiotics. The administration of γ-irradiated probiotics improved the clinical, biochemical, and histological parameters of colitis in TLR2$^{-/-}$ and TLR4$^{-/-}$ but not in TLR9$^{-/-}$ or MyD88$^{-/-}$ mice indicating the involvement of the TLR9 signaling pathway in the observed inhibition of colonic inflammation. In addition, administration of probiotic DNA or synthetic TLR9 ligand in the DSS and TNBS models as well as in spontaneous colitis in IL-10 deficient mice also inhibited colonic inflammation. Thus, in contrast to the current paradigm related to the proinflammatory role of TLR-activated innate immunity, our data indicated that TLR9 signaling results in the activation of an anti-inflammatory program that attenuates inflammation in different models of experimental colitis.

4   The Anti-Inflammatory Role of IFN1

In subsequent studies, we addressed the molecular basis for the anti-inflammatory effects induced by TLR9 signaling using the DSS model of experimental colitis. We found that two genetically distinct, but phenotypically similar, immune deficient mouse strains responded differently to TLR9 ligand administration.[8] While DSS-induced colitis was inhibited by TLR9 ligand in RAG$^{-/-}$ mice, colitis in SCID mice was not. We utilized these TLR9-responsive and TLR9-resistant phenotypes to dissect the anti-inflammatory role of TLR9 signaling in colonic inflammation. Analysis of the response to TLR9 ligand of these two mouse strains revealed defective TLR9-induced IFN1 production in SCID mice (most likely due to DNA-PK inactivation). Furthermore, we observed that IFNa/bR$^{-/-}$ mice are extremely susceptible to DSS-induced colitis and that the administration of TLR9 ligand to these mice increased the severity of the disease. Finally, the administration of recombinant IFNb (i.e., IFN1) to DSS-treated mice mimicked the anti-inflammatory effects on colonic inflammation induced by TLR9 ligand. Collectively, these sets of data indicate that IFN1 has a role in the maintenance of colonic homeostasis as well as in the amelioration of colonic injury.

5   The T Cell Independent Role of Dendritic Cells in Experimental Colitis

As dendritic cells (DCs) are the major cellular source of IFN1 induced by TLR signaling, we sought to uncover and define the role of these cells in colonic homeostasis.[1] To do so, we employed both depletion and adoptive transfer approaches. The depletion of cDCs (i.e., CD11c[hi]) by diphtheria toxin (DT) administration[7] to DSS-treated transgenic (tg)-DT receptor (DTR) mice notably inhibited experimental colitis (i.e., indicating a proinflammatory cDC phenotype), whereas the depletion of cDCs in TLR9 ligand/DSS-treated animals enhanced colonic inflammation (i.e., indicating an anti-inflammatory cDC phenotype). It is noteworthy that DT depletes neither plasmacytoid DCs nor macrophages in tg-DTR mice. To evaluate whether the results obtained for tg-DTR mice could be reproduced in a T cell independent fashion, we intercrossed tg-DTR to RAG$^{-/-}$ mice and then injected DT to DSS- and to TLR9 ligand/DSS-treated tg-DTR/RAG$^{-/-}$ mice under the same conditions used for tg-DTR mice. The depletion of cDCs in these immune deficient mice resulted in changes similar to those observed in the immune competent tg-DTR mice.

To test whether the two types of cDCs described above (i.e., pro- and anti-inflammatory cDC phenotypes) can transfer their regulatory functions to another animal, we adoptively transferred splenic-cDCs isolated from TLR9 ligand/DSS-treated wt mice to TLR9 ligand/DSS/DT-treated tg-DTR mice. Indeed, the transfer of splenic-cDCs isolated from TLR9 ligand/DSS-treated wt mice attenuated the colonic inflammation induced by cDC depletion in ISS/DT/DSS-treated tg-DTR mice. In contrast, when splenic-cDCs isolated from DSS-treated wt mice were adoptively transferred to DT/DSS-treated tg-DTR mice, they reversed the attenuation of colitis achieved upon cDC depletion.

Transcriptional analysis of cDCs isolated from TLR9-ligand/DSS-treated animals revealed high levels of IFN1 and IL-10 mRNAs in these cells. To further clarify which of these two cytokines inhibits colonic inflammation, rIFNb or rIL-10 was administered to TLR9 ligand/DT/DSS-treated tg-DTR mice. We hypothesized that the administration of these cytokines in the cDC-depleted animals could provide a clue as to which cDC-derived cytokine mediates the beneficial effect in this model. Indeed, systemic administration of recombinant IFNb, but not rIL-10, attenuated colonic inflammation. The inability of recombinant IL-10 to inhibit colonic inflammation is supported by our previous study demonstrating that administration of TLR9-ligand was effective in reducing colitis symptoms in IL-10$^{-/-}$ mice given DSS.

Since IFN1 regulates the expression of different chemokines, we hypothesized that IFN1 advances the resolution of inflammation by differential recruitment of neutrophils and monocytes into the inflamed colon. Indeed, IFNa/bR$^{-/-}$ mice displayed a significant delay in body weight gain 1 week after the withdrawal of DSS as compared to wt mice and reached their pre-DSS body weight 1 week after wt mice did. This observation correlated with the cellular infiltrates seen in the lamina propria (LP). While neutrophils were dominant in IFNa/bR$^{-/-}$ mice, F4/80-positive cells (i.e., macrophages) dominated the wt LP during the recovery phase. Consistent with these findings, KC, a neutrophil-attracting chemokine, was produced at significantly higher levels in the supernatants of colonic explants from IFNa/bR$^{-/-}$ than from wt mice, while RANTES, a monocyte/macrophage-attracting chemokine, was detected at higher levels in wt than in IFNa/bR$^{-/-}$ colonic explants.

Different types of hematopoietic cells contribute to colonic inflammation induced by DSS administration. However, whatever be the contribution of a particular cell type to this process, our results indicate that the extent and the severity of colonic inflammation are largely determined by cDCs. The depletion of cDCs in DSS-treated tg-DTR mice almost completely inhibited experimental colitis (i.e., indicates the proinflammatory phenotype of cDCs), whereas the depletion of cDCs in TLR9 ligand/DSS-treated animals enhanced colonic inflammation (i.e., indicates the anti-inflammatory phenotype of cDCs). This diverse regulatory role of cDCs is most likely the result of their special ability to generate, receive, integrate, and transmit a variety of biological signals. Due to their localization in the lamina propria, cDCs can affect the functionality of intestinal epithelial cells and stromal cells as well as blood-derived cells such as neutrophils and/or macrophages. In this respect, cDCs function as the central processor and are a key effector of colonic homeostasis in a T cell independent fashion. These results underline the pivotal function of cDCs in colonic homeostasis and define their role as the cellular switch that regulates at least this type of colonic inflammation. Depending on their mode of activation, cDCs (1) secrete a variety of inflammatory cytokines (e.g., TNFa, IL-12, IL-6), (2) secrete anti-inflammatory cytokines (e.g., IFN1) that inhibit both their own and macrophage produced inflammatory cytokines, (3) regulate the production of certain chemokines that dictate the composition of the cellular infiltrate (e.g., KC), and, consequently, (4) affect the pace of resolution of the inflamed colon. Collectively, these data expand the role of cDCs beyond T cell activation and position cDCs as important regulators of acute inflammation. We propose that intervention strategies designed to trigger an IFN1 response in the intestinal tract can have a therapeutic value for intestinal inflammatory conditions.

Acknowledgments This work was supported by NIH grants AI40682, AI68685, DK35108 and a grant from the CCFA. We thank Mrs. L. Beck for her editorial assistance.

References

1. Abe K, Nguyen KP, Fine SD, Mo J-H, Shen C, Shenouda S, Corr M, Jung S, Lee J, Eckmann L, Raz E (2007) Conventional dendritic cells regulate the outcome of colonic inflammation independently of T cells. Proc Natl Acad Sci USA (in press)

2. Akira S (2001) Toll-like receptors and innate immunity. Adv Immunol 78:1–56
3. Akira S, Hemmi H (2003) Recognition of pathogen-associated molecular patterns by TLR family. Immunol Lett 85:85–95
4. Dubinsky MC, Wang D, Picornell Y, Wrobel I, Katzir L, Quiros A, Dutridge D, Wahbeh G, Silber G, Bahar R, Mengesha E, Targan SR, Taylor KD, Rotter JI (2007) IL-23 receptor (IL-23R) gene protects against pediatric Crohn's disease. Inflamm Bowel Dis 13:511–515
5. Duerr RH, Taylor KD, Brant SR, Rioux JD, Silverberg MS, Daly MJ, Steinhart AH, Abraham C, Regueiro M, Griffiths A, Dassopoulos T, Bitton A, Yang H, Targan S, Datta LW, Kistner EO, Schumm LP, Lee AT, Gregersen PK, Barmada MM, Rotter JI, Nicolae DL, Cho JH (2006) A genome-wide association study identifies IL23R as an inflammatory bowel disease gene. Science 314:1461–1463
6. Hugot JP, Chamaillard M, Zouali H, Lesage S, Cezard JP, Belaiche J, Almer S, Tysk C, O'Morain CA, Gassull M, Binder V, Finkel Y, Cortot A, Modigliani R, Laurent-Puig P, Gower-Rousseau C, Macry J, Colombel JF, Sahbatou M, Thomas G (2001) Association of NOD2 leucine-rich repeat variants with susceptibility to Crohn's disease. Nature 411:599–603
7. Jung S, Unutmaz D, Wong P, Sano G, De los Santos K, Sparwasser T, Wu S, Vuthoori S, Ko K, Zavala F, Pamer EG, Littman DR, Lang RA (2002) In vivo depletion of CD11c(+) dendritic cells abrogates priming of CD8(+) T cells by exogenous cell-associated antigens. Immunity 17:211–220
8. Katakura K, Lee J, Rachmilewitz D, Li G, Eckmann L, Raz E (2005) Toll-like receptor 9-induced type I IFN protects mice from experimental colitis. J Clin Invest 115:695–702
9. Lee J, Mo JH, Katakura K, Alkalay I, Rucker AN, Liu YT, Lee HK, Shen C, Cojocaru G, Shenouda S, Kagnoff M, Eckmann L, Ben-Neriah Y, Raz E (2006) Maintenance of colonic homeostasis by distinctive apical TLR9 signalling in intestinal epithelial cells. Nat Cell Biol 8:1327–1336
10. Ogura Y, Bonen DK, Inohara N, Nicolae DL, Chen FF, Ramos R, Britton H, Moran T, Karaliuskas R, Duerr RH, Achkar JP, Brant SR, Bayless TM, Kirschner BS, Hanauer SB, Nunez G, Cho JH (2001) A frameshift mutation in NOD2 associated with susceptibility to Crohn's disease. Nature 411:603–606
11. Rachmilewitz D, Karmeli F, Takabayashi K, Hayashi T, Leider-Trejo L, Lee J, Leoni LM, Raz E (2002) Immunostimulatory DNA ameliorates experimental and spontaneous murine colitis. Gastroenterology 122:1428–1441
12. Rachmilewitz D, Katakura K, Karmeli F, Hayashi T, Reinus C, Rudensky B, Akira S, Takeda K, Lee J, Takabayashi K, Raz E (2004) Toll-like receptor 9 signaling mediates the anti-inflammatory effects of probiotics in murine experimental colitis. Gastroenterology 126:520–528
13. Watanabe T, Kitani A, Murray PJ, Wakatsuki Y, Fuss IJ, Strober W (2006) Nucleotide binding oligomerization domain 2 deficiency leads to dysregulated TLR2 signaling and induction of antigen-specific colitis. Immunity 25:473–485
14. Xavier RJ, Podolsky DK (2007) Unravelling the pathogenesis of inflammatory bowel disease. Nature 448:427–434

The NK/DC Complot

Emanuela Marcenaro, Mariella Della Chiesa, Silvia Pesce, Sophie Agaugué, and Alessandro Moretta(✉)

1 Introduction

Different NK cell subsets exist that display major functional differences in their cytolytic activity, cytokine production, and homing capabilities. In particular, CD56high CD16– NK cells, which largely predominate in lymph nodes, have little cytolytic activity but release high levels of cytokines; whereas CD56low CD16+ NK cells, which predominate in peripheral blood and inflamed tissues, display lower cytokine production but potent cytotoxicity.[1] Various cell types that are resident within peripheral tissues as well as circulating cells that have been recruited in response to chemokine gradients into inflamed sites are equipped with receptors for pathogen-associated products that induce cytokine release upon engagement by their specific ligands. These cytokines directly influence the ability of NK cells to modulate both innate and adaptive immune responses.[2] For example, innate cytokines such as IL12 and IL18, produced by antigen presenting cells (APCs) including monocyte-derived dendritic cells (DCs), by acting on NK cells at early stages of immune response, promote two distinct pathways of T cell priming each characterized by a sharp polarization toward Th1 priming.[3] On the contrary, exposure of NK cells to an IL4-rich milieu, resulting from the release of this cytokine by other innate immune cells such as mastocytes and eosinophils, leads to a deviation from Th1 responses toward nonpolarized T cell priming.[4] The polarizing effects of NK cells are exerted at two different stages: the first, taking place in peripheral inflamed tissues, is based on the "editing" process by which optimal maturation of DC is achieved,[5] while the second one takes place in secondary lymphoid tissues where NK cells upon release of IFN-gamma directly influence T cell polarization toward Th1 responses.[6]

Alessandro Moretta
Dipartimento di Medicina Sperimentale, Università degli Studi di Genova, Via L.B. Albertiz, 16132 Genova, Italy *and* Dipartimento di Medicina Sperimentale, Sezione di Istologia, Università degli Studi di Genova, Via G.B. Marsano 10, 16132 Genova, Italy

S.P. Schoenberger et al. (eds.) *Crossroads between Innate and Adaptive Immunity II*, doi: 10.1007/978-0-387-79311-5_2, © Springer Science + Business Media, LLC 2009

2   Natural Killer Cells Are Not So Natural

Natural killer cells originate from the bone marrow, circulate in the blood, and become activated within peripheral tissues by cytokines, pathogen-derived substances, or upon encountering target cells that express ligands for NK cell receptors.[7] These receptors display either activating or inhibitory function, and the balance of their signals determines the outcome of NK cell function. The main inhibitory receptors recognize HLA class I molecules, which are present on virtually all healthy cells, and prevent NK cell attack against these cells. Loss of HLA class I molecules from cells owing to infection or tumor transformation may lead to NK cell activation, as originally proposed by the "missing self hypothesis,"[8] provided that an activating receptor is engaged. The inhibitory receptors specific for HLA-class I molecules consist of two structurally distinct families of molecules: the killer cell Ig-like receptors (KIR) specific for allelic determinants of HLA-A, B, or C molecules and the killer cell lectin-like receptors (KLR) that include CD94/NKG2A specific for HLA-E.[9–12] Most activating NK receptors bind to host-derived or pathogen-encoded ligands that are upregulated on "stressed" or infected cells, while a minority may recognize HLA molecules.[13,14] Once activated, NK cells acquire regulatory capabilities that are based on the ability to release various cytokines upon engagement of different triggering NK receptors or upon a concerted signaling by other cytokine combinations. Remarkably, they also acquire the ability to directly lyse target cells by exocytosis/degranulation of perforin and granzymes. Degranulation by NK cells results in depletion of intracellular perforin and LAMP-1 (CD107a) appearance at the cell surface.[15] Long and coworkers[16–18] demonstrated that cytolytic granule polarization and degranulation are two steps in NK cell-mediated cytotoxicity that are controlled separately by signals emanating from distinct receptors. These studies demonstrated that neither polarization nor degranulation is sufficient for efficient target cell lysis; however, coengagement of activating receptors and of LFA-1 resulted in strong cytotoxicity. While in resting NK cells the only activating receptor inducing efficient cytotoxicity appears represented by CD16, in activated NK cells crosslinking of other activating receptors including NKp30 and NKp46 by plastic-bound specific mAbs is sufficient to induce CD107a exposure at the NK cell surface and a sharp decrement of the amount of intracellular perforin.[19] Since most functions mediated by NK cells are acquired only in inflamed tissues where these cells become activated, the above results suggest that circulating nonactivated NK cells may only mediate ADCC while all the remaining NK functions would be a direct consequence of their activation after extravasation.

3   The "Flip the Switch" Model

A series of soluble factors play an important role in the early events that favor the extravasation of NK cells and the subsequent induction of their priming. These include various cytokines and chemokines that are released by resident DC and other cell types including endothelial cells, macrophages, neutrophils, fibroblasts, mast cells, and eosinophils during pathogen-induced inflammation in peripheral tissues.[2] The mechanism of NK cell recruitment appears to involve chemokines such as CXCL8, CCL3, CX3CL1, and Chemerin.[20–22] Indeed most circulating NK cells (CD56low CD16+) express CXCR1, CX3CR1, and ChemR23, while the minor CD56high, CD16– NK subset expresses CCR7.[1,23,24] According to their chemokine receptor phenotype, CD56low CD16+ cells are mainly recruited in pathogen-invaded inflamed tissues, whereas CD56high, CD16– cells are essentially attracted to secondary lymphoid compartments such as lymph nodes. Migration toward lymph nodes is induced by CCL21, which is highly expressed in high endothelial venules (HEVs), lymphatic vessels, and stromal/interdigitating DC (predominantly in T cell areas), and by CCL19, which is highly expressed in mature DC within the T cell zone of paracortex.[25,26]

Thus, as expected, within normal noninflamed lymph nodes NK cells are homogeneously characterized by the CD56high, CD16– surface phenotype, by low levels of cytolytic activity, and by the production of high amounts of IFN-gamma.[26,27] Similar to the minor circulating subset expressing this phenotype also lymph node NK cells do not express KIR while their inhibitory receptor is represented by CD94/NKG2A. Interestingly, a recent report demonstrated that upon exposure to IL2, IL15, or IL12 in vitro peripheral blood CD56high NK cells may gain the typical pattern of CD56low NK cells including de novo KIR expression. Together with the observation that NK cells derived from reactive lymph nodes and follicular hyperplasia are often expressing a KIR+ phenotype, these data suggest that CD56high, CD16– lymph node NK cells exposed to cytokines such as IL2, released by T cells after priming, may switch their HLA class I receptor phenotype from KIR– to KIR+. In line with this concept KIR+ NK cells were detected in the efferent lymph.[28]

4 NK Cells Need to be Primed to Become Innate

In inflamed tissues NK cells were found in close proximity with both conventional monocyte-derived dendritic cells (DCs) and plasmacytoid DC (PDC),[22] thus further supporting the notion that this interaction will result in reciprocal cell activation. Remarkably, only a fraction of NK cells within such tissues express the early activation marker CD69, suggesting that only a fraction of the recruited NK cell population undergoes activation at inflammatory sites. As mentioned earlier, to undergo activation NK cells require not only extravasation but also priming within tissues. This priming requires not only signals delivered by triggering receptors but also cytokines released by DC. The orchestrated response of both NK cells and different types of DC is likely regulated by pattern recognition receptors such as TLRs. Indeed it has been shown that NK cells express TLR3 and TLR9 and that engagement of these receptors by the specific microbial ligands induces a first signal toward NK cell activation.[29] A second signal is simultaneously provided by conventional myeloid DC or PDC that upon triggering via TLR3 or TLR9 release IL12 and IFN-alpha, respectively.[29,30] In the presence of both signals, NK cells acquire the ability to release abundant cytokines and to kill various types of target cells including tumors and virus-infected cells. Along this line it is important to underline that circulating nonactivated NK cells display much lower cytolytic activity against a very restricted number of tumor cell lines. According to the majority of investigators, the only cell line susceptible to such NK cells is represented by the K562 erythroleukemia. Remarkably priming of NK cells in the presence of both TLR ligands and cytokines such as IL12 enables NK cells to kill immature DC (iDC) and to induce their maturation. This process (termed NK-mediated "editing")[5] appears to constitute an essential step for the generation of mature DC (mDC) that after migration into lymph nodes can polarize naïve T cells toward Th1 responses. Thus the process of NK cell priming within peripheral tissues not only generates the "innate" ability of NK cells to kill efficiently tumors and virus-infected cells but is also required for promoting an efficient NK/DC crosstalk characterized by a productive NK-mediated "editing" necessary for subsequent Th1 polarization. It should also be noted that the major NK cell population involved in the "editing" process is represented by a subset of CD56low CD16+ cells expressing CD94/NKG2A as a HLA class I-specific inhibitory receptor.[31] In this context it has also been shown that NK cells activated by tumors or viral-infected cells in vivo prime DC to produce IL12,[32,33] thereby promoting further NK cell priming and Th1 polarization. A process of NK cell priming has been recently demonstrated in mice by Lucas et al.[34] who showed that "naïve" NK cells do not acquire "effector" function unless a priming step has occurred by contact with DC in draining lymph nodes. Along this line Martin-Fontecha et al.[6] showed that NK cells within lymph nodes provide a first signal for subsequent Th1 polarization.

5 Innate Cytokines Direct and Modulate NK Cell Functions

While short-term exposure to APC-derived IL12 promotes the release by NK cells of high levels of both IFN-gamma and TNF-alpha and the acquisition of cytolytic activity, exposure to IL4 was shown to result in poor cytokine production and low cytolytic activity.[4] Accordingly, NK cells exposed to IL12 may favor the differentiation/selection of appropriate mature DC for subsequent Th1 cell priming in lymph nodes. On the contrary, NK cells exposed to IL4 would not exert DC selection, may impair efficient Th1 priming, and favor either tolerogenic or Th2-type responses ("Indolent NK cells"). A different situation is generated by the crosstalk between NK and PDC. In this case, in the presence of ligands for their TLR9, PDC release abundant IFN-alpha, while NK cells, stimulated via the same receptor, acquire the ability to upregulate upon direct cell-to-cell interaction the level of IFN-alpha release by PDC.[30] In turn this cytokine induces strong upregulation of NK cell-mediated killing. As a consequence NK cells are able to kill abnormal target cells including tumors. More importantly they acquire the ability to kill iDC and to promote their "editing" before they migrate into lymph nodes. Another recent study proposed that in addition to IL12, IL4, and IFN-alpha other cytokines, such as IL18, released by DC or by macrophages, in response to pathogens, may influence the "helper" activity of NK cells.[3] Thus IL18, but not other NK cell activating cytokines, would promote the development of a unique type of "helper" NK cells characterized by the CD56+/CD83+/CCR7+/CD25+ phenotype. These IL18-induced NK cells appear to display a distinctive ability to support IL12 production by DC and to promote Th1 responses by naïve CD4+ T cells (Fig. 1). Along this

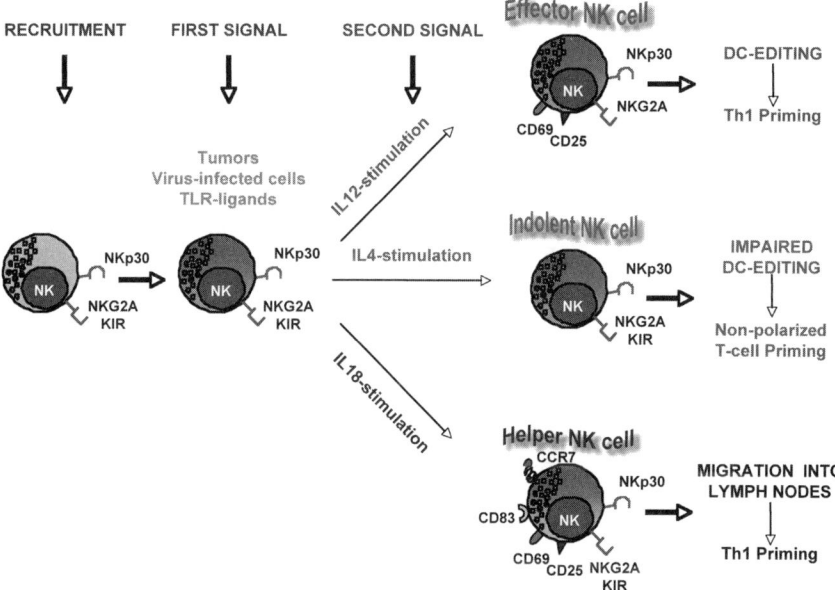

Fig. 1 Functionally divergent types NK cells are generated in inflamed peripheral tissues. CD56low CD16+ NK cells, after recruitment into peripheral inflammatory sites, become activated upon encounter with tumors or virus-infected cells that are recognized via different triggering receptors expressed at the NK surface. Alternatively NK cells may become activated upon encounter with pathogen-derived products that are recognized via TLRs ("first signal"). Once activated, NK cells acquire distinct functional characteristics depending on the prevalence of one or another cytokine released within the inflammatory microenvironment ("second signal"). In the presence of IL12 (released by DC upon Ag recognition) NK cells acquire their "effector" phenotype characterized by the ability to mediate the "editing" process. This is based on the ability to kill those DC that have not been suitably activated by the Ag. These DC, different from those undergoing appropriate maturation, do not upregulate HLA class I molecules and, similar to iDC, would remain susceptible to NK-mediated killing. In the meanwhile the same IL12-exposed "effector" NK cells release cytokines such as TNF-alpha and IFN-gamma that facilitate the progression of pathogen-responsive DC toward a full maturation. The latter is characterized by a strong upregulation of HLA expression, by the de novo expression of CCR7 and by the upregulation of costimulatory molecules such as CD80 and CD86. These mature DC migrate to secondary lymphoid compartments to promote Th1 polarization. Alternatively, upon exposure to IL18, NK cells may acquire a "helper" phenotype characterized by strong upregulation of IFN-gamma release but poor cytolytic activity. Different from NK cells carrying an "effector" phenotype, these NK cells migrate to secondary lymphoid organs upon acquisition of surface CCR7 (and CD83) expression that enables their response to CCL19/CCL21 cytokines. By contrast NK cells exposed to IL4 become "indolent" since they neither mediate an "editing" process nor migrate into lymph nodes. The outcome of their interaction with DC would be the promotion of nonpolarized T cell priming

line in a human in vitro system, using B lymphoma cells as a tumor model, it was also shown that the crosspresentation of cell-associated antigens to T cells by DC requires "help" from NK cells. These results suggest that capture of tumor cells and a full maturation status of DC are not sufficient to crossprime CD8+ T cells. Effective crosspriming would require further activation of DC by NK cells and an abundant production of IL18.[35]

6 Migratory NK Cells

The CCR7+ NK cell phenotype induced by IL18 (Fig. 2) is also characterized by high migratory responsiveness to lymph node-produced chemokines,[3] suggesting that not only "classical" CD56high, CD16– but also CD56low, CD16+ NK cells (at least under certain conditions) could migrate to lymph nodes. However CD56high, CD16– NK cells would reach these sites directly from the blood, whereas CD56low, CD16+ NK cells would follow a different route that includes their activation by IL18 released by APC encountered within inflamed peripheral tissues. Interestingly since IL18-induced "migratory" NK cells, unlike conventional CD56low, CD16+ NK cells, display low cytotoxicity but high cytokine release it is possible that their role within secondary lymphoid tissues may not be different from that of "resident" CD56high, CD16– NK cells. This finding however is important

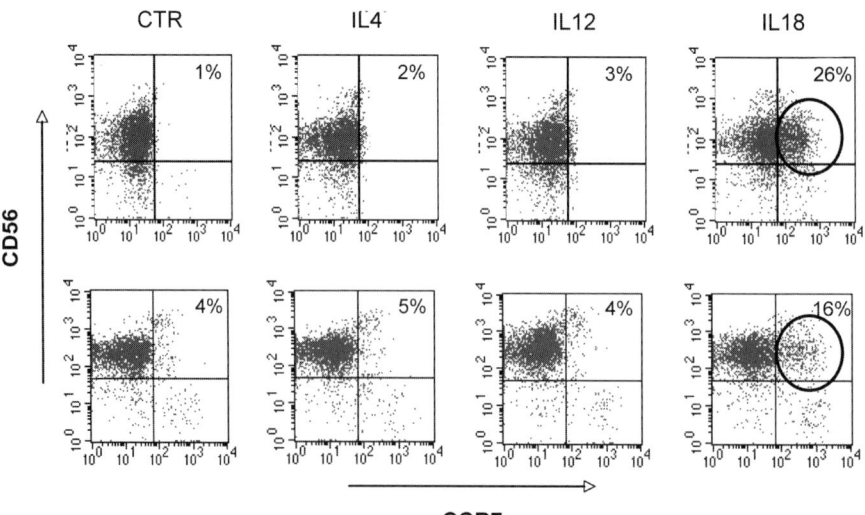

Fig. 2 Induction of CCR7 expression on the surface of human NK cells. Freshly purified NK cells from two donors (*upper and lower panels*) were cultured overnight in the presence of the indicated exogenous cytokines. Next cells were harvested and assessed for CD56 vs. CCR7 expression by cytofluorimetric analysis. *Circles* highlight the CD56+ CCR7+ population induced after culture in the presence of IL18

because it suggests that NK cells, similar to mDC, by migrating from inflamed tissues into regional lymph nodes, may directly deliver at these sites important signals contributing to T cell polarization. It will be important to clarify how other cytokines may impact this migratory capability and to understand whether the migratory phenotype is acquired by NK cells recruited at sites of inflammation independently from their phenotype. It is also possible that other still undefined signals may generate migratory NK cells independent from IL18 release.

7   Antigen Presentation by NK, NK–DC, or IKDC

Although DC are usually considered to be the most potent APC, the fact that activated NK cells express MHC class II, CD86, CD80, CD70, and OX40L strongly suggested the possibility that they might also communicate directly with CD4+ T cells.[36] Recent studies have demonstrated that indeed human NK cells efficiently enhance CD4+ as well as CD8+ T cell proliferation in response to antigen-specific stimulation or anti-CD3 treatment. Interestingly this process is dependent on direct contact-mediated interactions between ligands for T cell receptor (TCR), costimulatory receptors expressed on stimulated human NK cells (e.g., CD80, CD86, CD70, OX40 ligand, and 2B4 receptors) and their counterparts expressed on T cells (e.g., CD28, CD27, OX40, and CD48).[37–40] Moreover, due to the expression MHC class II, activated human NK cells have the capability to present antigens directly and stimulate CD4+ T cell proliferation in vitro.[39] Therefore, activated human NK cells possess not only the required costimulatory molecules for potential interaction with activated CD4+ T cells, but also have the capacity to process and present antigens through MHC class II. The term "presentation after killing" was used to indicate a process by which human NK cells utilize their cytotoxic features to kill target cells, acquire antigens from them, and present them to T cells in an antigen-specific manner.[39] The hypothesis for the presence of antigen presentation after killing in vivo was substantiated by the characterization of a unique bitypic NK–DC subset in mice that presents antigens to T cells after the killing of target cells. These cells were termed NK–DC or IFN-producing killer DC (IKDC).[41,42] These cells killed typical NK target cells using NK activating receptors, following which the DC-like antigen presenting activity was gained. Although expressing perforin and granzymes and producing substantial amounts of IFN-gamma, these cells were formally distinguished from classical NK cells. More recently, however, other studies[43–45] challenged the existence of IKDC by demonstrating that they express typical NK markers such as NKp46 and that their development is characterized by features typical of NK cells. Even the expression of MHC class II molecules, which was originally considered as one of the hallmarks of IKDC, was found to characterize not only these cells but also NK cells exposed to exogenous cytokines including IL15 and IL18. Thus one of these studies proposed that IKDC may simply represent a subset of activated NK cells, while another suggested that they may functionally represent the murine equivalent of human CD56high NK cells.

Acknowledgments This work was supported by grants awarded by Associazione Italiana per la Ricerca sul Cancro (A.I.R.C.), Istituto Superiore di Sanità (I.S.S.), Ministero della Salute – RF 2002/149, Ministero dell'Istruzione dell'Università e della Ricerca (M.I.U.R.), FIRB-MIUR progetto-cod.RBNE017B4C, Ministero dell'Università e della Ricerca Scientifica e Tecnologica (M.U.R.S.T.), European Union FP6, LSHB-CT-2004-503319-Allostem, and Compagnia di San Paolo.

References

1. Cooper, M.A., Fehniger, T.A. & Caligiuri, M.A. The biology of human natural killer-cell subsets. *Trends Immunol* **22,** 633–640 (2001)
2. Moretta, A. et al. Early liaisons between cells of the innate immune system in inflamed peripheral tissues. *Trends Immunol* **26,** 668–675 (2005)
3. Mailliard, R.B. et al. IL-18-induced CD83+CCR7+ NK helper cells. *J Exp Med* **202,** 941–953 (2005)
4. Marcenaro, E. et al. IL-12 or IL-4 prime human NK cells to mediate functionally divergent interactions with dendritic cells or tumors. *J Immunol* **174,** 3992–3998 (2005)
5. Moretta, A. Natural killer cells and dendritic cells: rendezvous in abused tissues. *Nat Rev Immunol* **2,** 957–964 (2002)
6. Martin-Fontecha, A. et al. Induced recruitment of NK cells to lymph nodes provides IFN-gamma for T(H)1 priming. *Nat Immunol* **5,** 1260–1265 (2004)
7. Moretta, A., Bottino, C., Mingari, M.C., Biassoni, R. & Moretta, L. What is a natural killer cell? *Nat Immuno* l**3,** 6–8 (2002)
8. Ljunggren, H.G. & Karre, K. In search of the 'missing self': MHC molecules and NK cell recognition. *Immunol Today* **11,** 237–244 (1990)
9. Moretta, A. et al. Receptors for HLA class-I molecules in human natural killer cells. *Annu Rev Immunol* **14,** 619–648 (1996)
10. Long, E.O. Regulation of immune responses through inhibitory receptors. *Annu Rev Immunol* **17,** 875–904 (1999)
11. Lopez-Botet, M., Llano, M., Navarro, F. & Bellon, T. NK cell recognition of non-classical HLA class I molecules. *Semin Immunol* **12,** 109–119 (2000)
12. Vilches, C. & Parham, P. KIR: diverse, rapidly evolving receptors of innate and adaptive immunity. *Annu Rev Immunol* **20,** 217–251 (2002)
13. Moretta, A. et al. Activating receptors and coreceptors involved in human natural killer cell-mediated cytolysis. *Annu Rev Immunol* **19,** 197–223 (2001)
14. Vivier, E., Nunes, J.A. & Vely, F. Natural killer cell signaling pathways. *Science* **306,** 1517–1519 (2004)
15. Bryceson, Y.T., March, M.E., Barber, D.F., Ljunggren, H.G. & Long, E.O. Cytolytic granule polarization and degranulation controlled by different receptors in resting NK cells. *J Exp Med* **202,** 1001–1012 (2005)
16. Bryceson, Y.T., March, M.E., Ljunggren, H.G. & Long, E.O. Activation, coactivation, and costimulation of resting human natural killer cells. *Immunol Rev* **214,** 73–91 (2006)
17. Bryceson, Y.T., March, M.E., Ljunggren, H.G. & Long, E.O. Synergy among receptors on resting NK cells for the activation of natural cytotoxicity and cytokine secretion. *Blood* **107,** 159–166 (2006)
18. Moretta, A., Biassoni, R., Bottino, C., Mingari, M.C. & Moretta, L. Natural cytotoxicity receptors that trigger human NK-cell-mediated cytolysis. *Immunol Today* **21,** 228–234 (2000)
19. Moretta, A., Marcenaro, E., Parolini, S., Ferlazzo, G. & Moretta, L. NK cells at the interface between innate and adaptive immunity. *Cell Death Differ* (2007)

20. Campbell, J.J. et al. Unique subpopulations of CD56+ NK and NK-T peripheral blood lymphocytes identified by chemokine receptor expression repertoire. *J Immunol* **166,** 6477–6482 (2001)
21. Robertson, M.J. Role of chemokines in the biology of natural killer cells. *J Leukocyte Biol* **71,** 173–183 (2002)
22. Parolini, S. et al. The role of chemerin in the colocalization of NK and dendritic cell subsets into inflamed tissues. *Blood* **109,** 3625–3632 (2007)
23. Vitale, M. et al. The small subset of CD56brightCD16– natural killer cells is selectively responsible for both cell proliferation and interferon-gamma production upon interaction with dendritic cells. *Eur J Immunol* **34,** 1715–1722 (2004)
24. Kawashima, D. et al. Augmented expression of secondary lymphoid tissue chemokine and EBI1 ligand chemokine in Crohn's disease. *J Clin Pathol* **58,** 1057–1063 (2005)
25. Ngo, V.N., Tang, H.L. & Cyster, J.G. Epstein–Barr virus-induced molecule 1 ligand chemokine is expressed by dendritic cells in lymphoid tissues and strongly attracts naive T cells and activated B cells. *J Exp Med* **188,** 181–191 (1998)
26. Ferlazzo, G. et al. The abundant NK cells in human secondary lymphoid tissues require activation to express killer cell Ig-like receptors and become cytolytic. *J Immunol* **172,** 1455–1462 (2004)
27. Fehniger, T.A. et al. CD56bright natural killer cells are present in human lymph nodes and are activated by T cell-derived IL-2: a potential new link between adaptive and innate immunity. *Blood* **101,** 3052–3057 (2003)
28. Romagnani, C. et al. CD56brightCD16– killer Ig-like receptor – NK cells display longer telomeres and acquire features of CD56dim NK cells upon activation. *J Immunol* **178,** 4947–4955 (2007)
29. Sivori, S. et al. CpG and double-stranded RNA trigger human NK cells by Toll-like receptors: induction of cytokine release and cytotoxicity against tumors and dendritic cells. *Proc Natl Acad Sci USA* **101,** 10116–10121 (2004)
30. Della Chiesa, M., Romagnani, C., Thiel, A., Moretta, L. & Moretta, A. Multidirectional interactions are bridging human NK cells with plasmacytoid and monocyte-derived dendritic cells during innate immune responses. *Blood* **108,** 3851–3858 (2006)
31. Della Chiesa, M. et al. The natural killer cell-mediated killing of autologous dendritic cells is confined to a cell subset expressing CD94/NKG2A, but lacking inhibitory killer Ig-like receptors. *Eur J Immunol* **33,** 1657–1666 (2003)
32. Mailliard, R.B. et al. Dendritic cells mediate NK cell help for Th1 and CTL responses: two-signal requirement for the induction of NK cell helper function. *J Immunol* **171,** 2366–2373 (2003)
33. Mocikat, R. et al. Natural killer cells activated by MHC class I(low) targets prime dendritic cells to induce protective CD8 T cell responses. *Immunity* **19,** 561–569 (2003)
34. Lucas, M., Schachterle, W., Oberle, K., Aichele, P. & Diefenbach, A. Dendritic cells prime natural killer cells by trans-presenting interleukin 15. *Immunity* **26,** 503–517 (2007)
35. Dao, T. et al. Natural killer cells license dendritic cell cross-presentation of B lymphoma cell-associated antigens. *Clin Cancer Res* **11,** 8763–8772 (2005)
36. Hanna, J. & Mandelboim, O. When killers become helpers. *Trends Immunol* **28,** 201–206 (2007)
37. Hanna, J. et al. Proteomic analysis of human natural killer cells: insights on new potential NK immune functions. *Mol Immunol* **42,** 425–431 (2005)
38. Zingoni, A. et al. Cross-talk between activated human NK cells and CD4+ T cells via OX40–OX40 ligand interactions. *J Immunol* **173,** 3716–3724 (2004)
39. Hanna, J. et al. Novel APC-like properties of human NK cells directly regulate T cell activation. *J Clin Invest* **114,** 1612–1623 (2004)
40. Hanna, J. et al. Novel insights on human NK cells' immunological modalities revealed by gene expression profiling. *J Immunol* **173,** 6547–6563 (2004)
41. Taieb, J. et al. A novel dendritic cell subset involved in tumor immunosurveillance. *Nat Med* **12,** 214–219 (2006)

42. Chan, C.W. et al. Interferon-producing killer dendritic cells provide a link between innate and adaptive immunity. *Nat Med* **12,** 207–213 (2006)
43. Blasius, A.L., Barchet, W., Cella, M. & Colonna, M. Development and function of murine B220+CD11c+NK1.1+ cells identify them as a subset of NK cells. *J Exp Med* **204,** 2561–2568 (2007)
44. Vosshenrich, C.A. et al. CD11cloB220+ interferon-producing killer dendritic cells are activated natural killer cells. *J Exp Med* **204,** 2569–2578 (2007)
45. Caminschi, I. et al. Putative IKDCs are functionally and developmentally similar to natural killer cells, but not to dendritic cells. *J Exp Med* **204,** 2579–2590 (2007)

Detection of Microbes by Natural Killer T Cells

Yuki Kinjo and Mitchell Kronenberg(✉)

1   Introduction

Natural killer T (NKT) cells combine features of the innate and adaptive immune systems. For example, they are lymphocytes that express an $\alpha\beta$ T cell antigen receptor (TCR), typical of adaptive immunity, but they also express NK receptors, such as NK1.1 (NKR-P1 or CD161c), similar to NK cells, which are part of the innate immune system.[1,2] In mice, the majority of NKT cells express an invariant (*i*) TCRα chain with Vα14–Jα18 rearrangement.[1,2] We refer to these lymphocytes here as Vα14*i*NKT cells. These cells have a limited repertoire of TCRβ chains, mainly Vβ8.2, Vβ7, and Vβ2, with the highest representation of Vβ8.2 (more than 50%). Humans have a similar population that mostly expresses an invariant Vα24–Jα18 rearrangement with Vβ11 (Vα24*i*NKT).[1,2] We refer to these two populations in mice and humans as *i*NKT cells.

*i*NKT cells have several unique features. In contrast to conventional T cells that recognize peptide antigens presented by major histocompatibility complex (MHC) class I or class II molecules, *i*NKT cells recognize lipid antigens presented by CD1d, a nonclassical class I like antigen-presenting molecule. CD1d has a deep, narrow, and hydrophobic antigen-binding groove that is adapted for the presentation of lipid antigens, which are mostly glycolipids.[3] The antigen that has been used most and that is most efficient for stimulating *i*NKT cells is α-galactosylceramide (αGalCer), a synthetic glycosphingolipid (GSL) which was originally isolated from a marine sponge.[1,2] It is believed that the α linkage of sugar to the ceramide lipid is critical for the recognition by the invariant TCR, because β-linked GSLs, which are abundant components of the mammalian body, are not usually antigenic.

In mice, Vα14*i*NKT cells are CD4 positive or CD4, CD8 double negative (DN), and they are most abundant in liver, where these cells constitute 10–40% of

Mitchell Kronenberg
La Jolla Institute for Allergy and Immunology, 9420 Athena Circle, La Jolla, CA 92037, USA

S.P. Schoenberger et al. (eds.) *Crossroads between Innate and Adaptive Immunity II*, doi: 10.1007/978-0-387-79311-5_3, © Springer Science+Business Media, LLC 2009 lymphocytes. They are also present in many other organs and tissues such as spleen, thymus, bone marrow, blood, and lung with lower frequency (1% or less).[1,2] In humans, Vα24iNKT cells are less abundant than in mice, and while they are mostly CD4 or DN, a minority is CD8 positive. iNKT cells express CD44 and CD69, which are markers of activated or memory T cells. Consistent with the activated or memory phenotype, these cells produce a large amount of cytokines, including IFNγ and IL-4, very rapidly (1–2 h) after TCR stimulation by lipid antigens presented by CD1d.[1,2] In mice, iNKT cells constitutively express mRNA for these cytokines even before activation,[4,5] which may help to explain why these cells can secrete cytokines so rapidly upon TCR stimulation. Acquisition of cytokine transcripts and activation of marker expression occur in the thymus, and therefore these cells are sometimes referred to as natural memory cells, because the memory/ activated cell phenotype and behavior result from their developmental pathway in the thymus. Furthermore, like innate immune cells, not only do iNKT cells respond very quickly, but also there is so far no evidence for a memory response by these cells following antigen activation.

Once activated, iNKT cells stimulate many types of cells such as antigen-presenting cells (APC), NK cells, and conventional T and B lymphocytes.[1,6] Because of these unique features, iNKT cells have been implicated in various immune responses including the maintenance of tolerance and the response to tumors and infectious agents. In this chapter, we describe the general mechanisms as to how iNKT cells are activated by microorganisms.

2   Role of Vα14iNKT Cells in Host Defense Against Pathogens

It has been shown that Vα14iNKT cells play a role in bacterial clearance in infection with certain bacteria including *Streptococcus pneumoniae*[7] and *Pseudomonas aeruginosa*.[8] Joyee et al. recently observed that Jα18-deficient mice, which lack Vα14iNKT cells because they cannot form the invariant α gene rearrangement,[9] had increased susceptibility to *Chlamydia pneumoniae* infection.[10] Following intranasal infection with *C. pneumoniae*, Jα18-deficient mice exhibited higher bacterial number and more severe tissue damage due to excessive inflammation in lung compared to wild-type mice. Vα14iNKT cells produced IFNγ after *C. pneumoniae* infection, and they contributed to the induction of IFNγ production by conventional CD4 and CD8 T cells.[10] These results suggest that Vα14iNKT cells play a role in the response to *C. pneumoniae* through the induction of IFNγ production. Interestingly, however, Vα14iNKT cells play a suppressive role in host defense against a different *Chlamydia* species, *C. muridarum*, because Jα18-deficient mice are more resistant to *C. muridarum* infection through the decreased production of IL-4.[10] This study illustrates the dual role of Vα14iNKT cells. In some cases, these cells can become protective to the host by augmenting antimicrobial responses, but in other cases, the increased cytokine production by these cells is detrimental to the host.[11]

Vα14*i*NKT cells also play a role in the response to several parasites including *Leishmania major*[12,13] and *L. donovani*.[14] Duthie et al.[15] have found that Vα14*i*NKT cells regulate the inflammation after infection with *Trypanosoma cruzi*, the causative agent of Chagas disease. *J*α*18*-deficient mice showed very severe inflammation in the liver, spleen, and muscle after *T. cruzi* infection. Consistent with this result, increased production of inflammatory cytokines and accumulation of activated leukocytes in liver and spleen of *J*α*18*-deficient mice were observed. Interestingly, *Cd1d*-deficient mice were not susceptible to *T. cruzi* infection, suggesting that CD1d reactive cells that do not have Vα14–*J*α18 rearrangement are responsible for the severe inflammation during *T. cruzi* infection, as opposed to the anti-inflammatory response by Vα14*i*NKT cells.[15]

Vα14*i*NKT cells even participate in the response to viruses, although enveloped viruses generally only contain host glycolipids. It was reported that *Cd1d*-deficient mice were susceptible to infection with herpes simplex virus (HSV) types 1[16] and 2.[17] It was also shown that CD1d expression was downregulated by several viruses including vaccinia virus,[18] lymphocytic choriomeningitis virus,[19] Kaposi's sarcoma-associated herpes virus,[20] HSV[21] and HIV.[22] The downregulation of CD1d expression suggests a viral immune evasion mechanism to prevent a response by *i*NKT cells, or perhaps by other CD1d reactive T lymphocytes.

3   Vα14*i*NKT Cells Augment Innate Immune Responses Through the Production of IFNγ

Following earlier studies suggesting that mice lacking *i*NKT cells were resistant to endotoxin-induced shock,[23] Nagarajan et al.[24] observed that *J*α*18*-deficient mice, which lack Vα14*i*NKT cells, exhibited diminished serum TNF after LPS injection. Consistent with this observation, TNF production by neutrophils and macrophages was decreased in *J*α*18*-deficient mice after LPS injection. Treatment of *J*α*18*-deficient mice with IFNγ partially recovered serum TNF when LPS was coadministrated.[24] Furthermore, IFNγ production by NK cells in the liver and spleen was also impaired after LPS injection in *J*α*18*-deficient mice. These data suggest that early IFNγ production by Vα14*i*NKT cells augments the innate immune response to LPS through the stimulation of IFNγ production by NK cells and TNF production by neutrophils and macrophages.

Similarly, Vα14*i*NKT cells play an important role in the augmentation of the innate immune response against *S. pneumoniae*. *J*α*18*-deficient mice were susceptible to lung infection with *S. pneumoniae*,[7] and Nakamatsu et al.[25] recently reported that the administration of recombinant IFNγ could rescue *J*α*18*-deficient mice by recruiting neutrophils to the lung and by inducing production of MIP-2 and TNF there. Similarly, increased resistance to *S. pneumoniae* infection was observed in *J*α*18*-deficient mice that received liver mononuclear cells (MNC) from WT mice, containing approximately 25% Vα14*i*NKT cells.[25] However, the transfer of liver MNC from *J*α*18*-deficient mice or IFNγ-deficient mice into *J*α*18*-deficient mice failed to induce the production of MIP-2 and TNF. Neutrophil recruitment was also impaired, and the mice could not eliminate *S. pneumoniae* from the lung. These data suggest that Vα14*i*NKT cells stimulate the innate immune response to *S. pneumoniae* through the production of IFNγ.

4   Detection of Microbes by *i*NKT Cells

*i*NKT cells play a role in the response to various microorganisms; however, it remained elusive how these cells are activated during microbial infection, especially in cases where foreign antigens for the invariant TCR were not present. Brigl et al. proposed a mechanism as to how *i*NKT cells are activated in such situations.[26] They observed that *i*NKT cells produced IFNγ after *Salmonella* infection, although analysis of the lipid fraction extracted from these bacteria indicated that they do not contain lipid antigens for the invariant TCR. They found instead that IL-12 derived from LPS-stimulated dendritic cells (DC) was required for this response. Furthermore, blocking of the TCR/CD1d interaction with an anti-CD1d monoclonal antibody (mAb) could inhibit IFNγ production by the *i*NKT cells. These data suggest that both the recognition of endogenous ligand presented by CD1d and IL-12 are required for *i*NKT cell activation during *Salmonella* infection. Another group reported similar data, and they showed that MyD88 expression by the DC was required for *i*NKT cell activation.[27]

Nagarajan et al.[24] recently reported that IL-12 and IL-18 produced by LPS-stimulated DC are necessary and sufficient for IFNγ production by Vα14*i*NKT cells. They observed IFNγ production by Vα14*i*NKT cells after injection of *E. coli* LPS. By using IL-12p40-deficient mice and IL-18-deficient mice, it was shown that IL-12 and IL-18 were required for this Vα14*i*NKT cell response in vivo. When purified Vα14*i*NKT cells were cultured with wild-type DC, LPS could induce IFNγ production by these cells. CD1d-deficient DC stimulated with *E. coli* LPS could still induce *i*NKT cell IFNγ, showing that CD1d antigen recognition is not absolutely required. Furthermore, purified Vα14*i*NKT cells could make IFNγ when these cells were cultured with IL-12 and IL-18, even without DC. These data indicate that the inflammatory cytokines derived from LPS-stimulated DC are sufficient for inducing IFNγ production by Vα14*i*NKT cells, although in some cases, such as in the experiments where *Salmonella* or LPS from *Salmonella* were the stimulating agents, recognition of self-antigens presented by CD1d may also play a role. Finally, in the innate-like response of *i*NKT cells to Schistosomal extracts, self-antigen recognition was important, but IL-12 secretion was not.[28]

What unites all these antimicrobial responses is the absence of a microbial antigen for the invariant TCR, but it remains to be determined why the relative importance of cytokines from innate cells vs. self-antigen recognition varies in the different cases. It has been proposed that the GSL isoglobotrihexosylceramide (iGb3) is the dominant endogenous antigen for *i*NKT cells, required for their development and for their activation in the periphery.[29] Several recent studies contradict this finding,[30–32] however, and clearly the issue of self-antigen recognition cannot be fully resolved until the self-antigens are better defined.

5   Microbial Antigens for the Invariant TCR of iNKT Cells

It has been controversial as to whether iNKT cells could recognize microbial glycolipid antigens by their invariant TCR. An early study suggested that IL-4 producing NK1.1[+]T cells recognize the glycophosphatidylinositol (GPI) from *Plasmodium* and *Trypanosoma*, contributing to IgG reactive to GPI.[33] However, this finding could not be repeated by other groups.[34,35]

Fischer et al.[36] reported that a purified phosphatidyl inositol mannoside (PIM) extracted from *Mycobacterium bovis* BCG and enriched for PI tetramannosides (PIM$_4$), could bind to CD1d and stimulate T cells from Vα14 transgenic mice and Vα24iNKT cells. CD1d tetramers loaded with PIM could detect only a small minority of αGalCer reactive cells, however, especially in mouse liver. These results suggest that PIM$_4$ reactive cells are only a subset of iNKT cells.

Amprey et al.[14] observed that several percentages of Vα14iNKT cells in liver produced IFNγ 2 h after *Leishmania donovani* infection, to which CD1d-deficient mice have increased susceptibility. IFNγ production induced by *L. donovani* was IL-12 independent and CD1d dependent. They found that a lipophosphoglycan (LPG) purified from *Leishmania* could inhibit αGalCer-induced cytokine production by iNKT cell hybridomas using plates coated with CD1d to stimulate the hybridomas. This suggests that LPG can bind to CD1d. LPG-pulsed DC stimulated IFNγ production by liver MNC in a CD1d-dependent manner. Furthermore, LPG could induce IFNγ production by Vα14iNKT cells in liver 2 h after injection, although only 1.4% of αGalCer tetramer positive cells produced this cytokine.

These studies provided evidence that iNKT cells could recognize microbial glycolipid antigens using their invariant TCR, although only a minor subset of the iNKT cells reacted to any of the antigens described above.

6   Recognition of *Sphingomonas* GSLs by iNKT Cells

Are there any microbial antigens that can stimulate the majority of iNKT cells? We and other two groups found that GSLs from *Sphingomonas* spp. could stimulate the majority of iNKT cells.[27,37,38] *Sphingomonas* are Gram-negative bacteria that lack LPS, and they are abundant in the environment.[39] Instead of LPS, these bacteria have GSLs that contain monosaccharide sugars, either galacturonic acid (GalA)[40] or a glucuronic acid (GlcA).[41] In all cases, the sugars are α-linked to the lipid, similar to αGalCer. The entire structure of the prototypical *Sphingomonas* GSL is very similar to αGalCer, and in addition to the carboxylate modification at the 6 position of the galactose sugar, they include only a shorter acyl chain in the ceramide lipid, C14 in

GSL as opposed to C18 in αGalCer, and the lack of a hydroxyl group in the C4 position of sphingosine that is found in αGalCer.[11,37] The results from several in vitro and in vivo assays showed that purified and synthetic GalA-GSL and GlcA-GSL could bind to CD1d and stimulate Vα14iNKT cells and Vα24iNKT cells.[27,37,38] CD1d tetramers loaded with either GalA-GSL or GlcA-GSL could detect all the Vα24iNKT cells in expanded cell lines from peripheral blood, and at least half of the Vα14iNKT cells analyzed directly ex vivo. The reactive cells were absent in *Cd1d*-deficient mice and *Jα18*-deficient mice, showing that *Sphingomonas* GSL reactive cells and αGalCer reactive cells were overlapping. In contrast to the response to *Salmonella*, TLR signaling mediated by MyD88 or Trif and IL-12 was not required for the *Sphingomonas* GSL mediated in vitro or in vivo activation of Vα14iNKT cells.[27,37] These data suggest that iNKT cells can be activated directly by the recognition of microbial antigens by their invariant TCR, as opposed to indirect activation by inflammatory cytokines produced by TLR-stimulated APC.

Cd1d-deficient mice and *Jα18*-deficient mice showed delayed bacterial clearance from several sites, including the liver and the lung, after infection with *Sphingomonas* spp., suggesting that iNKT cell activated by the recognition of bacterial glycolipid antigens contributes to bacterial clearance.[27,37] However, when a high dose of *Sphingomonas* spp. was used, the control mice died from shock, whereas most iNKT cell-deficient mice survived.[27] This result suggests that the overactive response of iNKT cells in mice can be detrimental to the host. *Sphingomonas* bacteria are not highly pathogenic in humans, although infections in immunocompromised patients have been reported.[42,43] Because α-linked GSL can be found only in *Sphingomonas* spp., it remained unknown if iNKT cells can recognize other classes of glycolipids, especially those from pathogenic microbes.

7 iNKT Cells Recognize Galactosyl Diacylglycerols from *Borrelia burgdorferi*

Borrelia burgdorferi also lacks LPS. This spirochete is the causative agent of Lyme disease, which is the most common vector-borne disease in the USA. It was previously reported that *Cd1d*-deficient mice had increased bacterial number and increased joint swelling after *B. burgdorferi* infection,[44] suggesting a role for iNKT cells, and we observed activation of iNKT cells in liver after *B. burgdorferi* infection.

It has been previously shown that *B. burgdorferi* possesses two glycolipids, BbGL-I and BbGL-II, which comprise 23 and 12% of total lipid, respectively.[45] BbGL-1 is a cholesteryl β-galactoside, and BbGL-II is a galactosyl diacylglycerol. Although the glycerol lipid has a different structure from the ceramide lipid in GSLs, the α linkage of the galactose sugar suggested to us that BbGL-II could be an antigen for iNKT cells. In fact, we did find that BbGL-II extracted from *B. burgdorferi* could stimulate Vα14iNKT cell hybridomas in the CD1d-coated plate assay.[46] Purified BbGL-II contains a mixture of C14:0, C16:0, C18:0, C18:1, and C18:2

fatty acids, with C16:0 and C18:1 being the most abundant.[45] Based on this, we synthesized several BbGL-II compounds that have the different combination of these fatty acid chains. Although most of these synthetic compounds stimulated several Vα14*i*NKT cell hybridomas in the CD1d-coated plate assay, BbGL-IIc that has an oleic acid (C18:1) in the *sn-1* position of the glycerol and a palmitic acid (C16:0) in the *sn-2* position induced the most potent stimulation among these compounds.[46] This finding was confirmed by another study.[47] CD1d tetramers loaded with BbGL-IIc could detect approximately 25% of the Vα14*i*NKT cells in liver.[46] This may be an underestimate of the size of the reactive population, because all the Vα14*i*NKT cell hybridomas tested could react to BbGL-IIc. DC pulsed with BbGL-IIc could stimulate IFNγ production by Vα14*i*NKT cells in vivo in a TLR-independent manner. These data show that a galactosyl diacylglycerol from *B. burgdorferi* can stimulate many Vα14*i*NKT cells without indirect activation of APC. BbGL-II compounds also induced the production of IFNγ and IL-4 from expanded lines of human Vα24*i*NKT cells.[46] However, the response pattern of Vα24*i*NKT cells was different from that of Vα14*i*NKT cells. Vα24*i*NKT cells responded to the BbGL-II compounds that have a higher degree of unsaturation in their fatty acids, but not to BbGL-IIc.[46] This difference in the response to diacylglycerol compounds between human Vα24*i*NKT cells and mouse Vα14*i*NKT cells was striking. It is possible that the mouse and human CD1d grooves have very different preferences for CD1d binding. Alternatively, the nature of the aliphatic chains could influence the position of the protruding hydrophilic head group, with the invariant TCRs of the mouse and human having recognized subtly different epitopes. The influence of subtle changes in the lipid moiety upon antigenic potency is most surprising, and further studies will be needed to address the mechanism underlying this high degree of discrimination.

The results from the study of the *B. burgdorferi* response have several important implications. First, they establish that broad populations, likely the majority of *i*NKT cells, can recognize glycolipid antigens from a pathogenic microorganism. This finding suggests that these cells can detect pathogens using their invariant TCR and that they contribute to the response against these pathogens. Second, while GSLs are confined to *Sphingomonas*, similar diacylglycerol glycolipids have been found in other microorganisms. *i*NKT cells therefore could have reactivity to a variety of pathogens, although it must be emphasized that some microbes, such as *Salmonella*, apparently lack such antigens. Third, we speculate that microorganisms could evade the immune response mediated by *i*NKT cells by producing glycolipids that differ in the degree of saturation or the length of the fatty acids.

8   Conclusions

Recent studies emphasize the importance of *i*NKT cells in the host response to diverse microorganisms, and they shed light on the mechanism for this response. *i*NKT cells can detect pathogens such as *Salmonella* that do not contain lipid antigens for the invariant TCR. Their activation is initiated by the stimulation of APC, resulting in the production of inflammatory cytokines. This indirect activation mechanism confers upon *i*NKT cells the ability to respond to a variety of pathogens, including viruses. In addition to cytokines such as IL-12 and IL-18, it was shown that the simultaneous recognition of endogenous antigens presented by CD1d is important in some cases. Future studies are needed to define the relevant self-antigens. *i*NKT cells also can detect *Sphingomonas* spp. and *B. burgdorferi* by the recognition of glycolipids by their invariant TCR. Considering the wider distribution of the glycerol-containing glycolipids, we speculate that *i*NKT cells will play an important role in the response to additional pathogens, especially those that have diacylglycerol antigens. A recent study by Scott-Browne et al.,[48] together with the solved structure of a Vα24*i* TCR/αGalCer/human CD1d complex, provides a mechanistic explanation as to how the Vα14*i*NKT cell TCR can detect a variety of α-linked glycolipids. They found that "hot spot" of amino acids in the TCR complementarity-determining regions (CDRs), particularly those in CDR1 and CDR3 of the α chain, were almost same for the recognition of structurally distinct antigens.[48] This suggests that the invariant TCR of *i*NKT cells can detect a variety of antigens from various pathogens, acting almost as a type of pattern recognition receptor for the α-linked hexose sugars of glycolipid antigens.

Acknowledgments This work was supported by NIH grants R37 AI71922, RO1 AI45053, RO1 AI69276 (MK), and a fellowship from The Irvington Institute Fellowship Program of the Cancer Research Institute Research (Y.K.).

References

1. Kronenberg, M. Toward an understanding of NKT cell biology: progress and paradoxes. *Annu Rev Immunol* **23**, 877–900 (2005)
2. Godfrey, D.I. & Berzins, S.P. Control points in NKT-cell development. *Nat Rev Immunol* **7**, 505–518 (2007)
3. Moody, D.B., Zajonc, D.M. & Wilson, I.A. Anatomy of CD1-lipid antigen complexes. *Nat Rev Immunol* **5**, 387–399 (2005)
4. Stetson, D.B. . Constitutive cytokine mRNAs mark natural killer (NK) and NK T cells poised for rapid effector function. *J Exp Med* **198**, 1069–1076 (2003)
5. Matsuda, J.L. et al. Mouse V alpha 14i natural killer T cells are resistant to cytokine polarization in vivo. *Proc Natl Acad Sci USA* **100**, 8395–8400 (2003)
6. Brigl, M. & Brenner, M.B. CD1: antigen presentation and T cell function. *Annu Rev Immunol* **22**, 817–890 (2004)
7. Kawakami, K. Critical role of Valpha14+natural killer T cells in the innate phase of host protection against *Streptococcus pneumoniae* infection. *Eur J Immunol* **33**, 3322–3330 (2003)
8. Nieuwenhuis, E.E. CD1d-dependent macrophage-mediated clearance of *Pseudomonas aeruginosa* from lung. *Nat Med* **8**, 588–593 (2002)
9. Cui, J. Requirement for Valpha14 NKT cells in IL-12-mediated rejection of tumors. *Science* **278**, 1623–1626 (1997)
10. Joyee, A.G. . Distinct NKT cell subsets are induced by different *Chlamydia* species leading to differential adaptive immunity and host resistance to the infections. *J Immunol* **178**, 1048–1058 (2007)

11. Tupin, E., Kinjo, Y. & Kronenberg, M. The unique role of natural killer T cells in the response to microorganisms. *Nat Rev Microbiol* **5**, 405–417 (2007)

12. Ishikawa, H. CD4(+) v(alpha)14 NKT cells play a crucial role in an early stage of protective immunity against infection with *Leishmania major*. *Int Immunol* **12**, 1267–1274 (2000)

13. Mattner, J., Donhauser, N., Werner-Felmayer, G. & Bogdan, C. NKT cells mediate organ-specific resistance against *Leishmania major* infection. *Microbes Infect* **8**, 354–362 (2006)

14. Amprey, J.L. A subset of liver NK T cells is activated during *Leishmania donovani* infection by CD1d-bound lipophosphoglycan. *J Exp Med* **200**, 895–904 (2004)

15. Duthie, M.S., Kahn, M., White, M., Kapur, R.P. & Kahn, S.J. Critical proinflammatory and anti-inflammatory functions of different subsets of CD1d-restricted natural killer T cells during *Trypanosoma cruzi* infection. *Infect Immun* **73**, 181–192 (2005)

16. Grubor-Bauk, B., Simmons, A., Mayrhofer, G. & Speck, P.G. Impaired clearance of herpes simplex virus type 1 from mice lacking CD1d or NKT cells expressing the semivariant V alpha 14-J alpha 281 TCR. *J Immunol* **170**, 1430–1434 (2003)

17. Ashkar, A.A. & Rosenthal, K.L. Interleukin-15 and natural killer and NKT cells play a critical role in innate protection against genital herpes simplex virus type 2 infection. *J Virol* **77**, 10168–10171 (2003)

18. Renukaradhya, G.J. . Virus-induced inhibition of CD1d1-mediated antigen presentation: reciprocal regulation by p38 and ERK. *J Immunol* **175**, 4301–4308 (2005)

19. Lin, Y., Roberts, T.J., Spence, P.M. & Brutkiewicz, R.R. Reduction in CD1d expression on dendritic cells and macrophages by an acute virus infection. *J Leukocyte Biol* **77**, 151–158 (2005)

20. Sanchez, D.J., Gumperz, J.E. & Ganem, D. Regulation of CD1d expression and function by a herpesvirus infection. *J Clin Invest* **115**, 1369–1378 (2005)

21. Yuan, W., Dasgupta, A. & Cresswell, P. Herpes simplex virus evades natural killer T cell recognition by suppressing CD1d recycling. *Nat Immunol* **7**, 835–842 (2006)

22. Chen, N. HIV-1 down-regulates the expression of CD1d via Nef. *Eur J Immunol* **36**, 278–286 (2006)

23. Dieli, F. Resistance of natural killer T cell-deficient mice to systemic Shwartzman reaction. *J Exp Med* **192**, 1645–1652 (2000)

24. Nagarajan, N.A. & Kronenberg, M. Invariant NKT cells amplify the innate immune response to lipopolysaccharide. *J Immunol* **178**, 2706–2713 (2007)

25. Nakamatsu, M. Role of interferon-gamma in Valpha14+natural killer T cell-mediated host defense against *Streptococcus pneumoniae* infection in murine lungs. *Microbes Infect* **9**, 364–374 (2007)

26. Brigl, M., Bry, L., Kent, S.C., Gumperz, J.E. & Brenner, M.B. Mechanism of CD1d-restricted natural killer T cell activation during microbial infection. *Nat Immunol* **4**, 1230–1237 (2003)

27. Mattner, J. Exogenous and endogenous glycolipid antigens activate NKT cells during microbial infections. *Nature* **434**, 525–529 (2005)

28. Mallevaey, T. Activation of invariant NKT cells by the helminth parasite schistosoma mansoni. *J Immunol* **176**, 2476–2485 (2006)

29. Zhou, D. Lysosomal glycosphingolipid recognition by NKT cells. *Science* **306**, 1786–1789 (2004)

30. Gadola, S.D. Impaired selection of invariant natural killer T cells in diverse mouse models of glycosphingolipid lysosomal storage diseases. *J Exp Med* **203**, 2293–2303 (2006)

31. Porubsky, S. Normal development and function of invariant natural killer T cells in mice with isoglobotrihexosylceramide (iGb3) deficiency. *Proc Natl Acad Sci USA* **104**, 5977–5982 (2007)

32. Speak, A.O. Implications for invariant natural killer T cell ligands due to the restricted presence of isoglobotrihexosylceramide in mammals. *Proc Natl Acad Sci USA* **104**, 5971–5976 (2007)

33. Schofield, L. CD1d-restricted immunoglobulin G formation to GPI-anchored antigens mediated by NKT cells. *Science* **283**, 225–229 (1999)

34. Molano, A. Cutting edge: the IgG response to the circumsporozoite protein is MHC class II-dependent and CD1d-independent: exploring the role of GPIs in NK T cell activation and antimalarial responses. *J Immunol* **164**, 5005–5009 (2000)

35. Romero, J.F., Eberl, G., MacDonald, H.R. & Corradin, G. CD1d-restricted NK T cells are dispensable for specific antibody responses and protective immunity against liver stage malaria infection in mice. *Parasite Immunol* **23**, 267–269 (2001)

36. Fischer, K. Mycobacterial phosphatidylinositol mannoside is a natural antigen for CD1d-restricted T cells. *Proc Natl Acad Sci USA* **101**, 10685–10690 (2004)

37. Kinjo, Y. Recognition of bacterial glycosphingolipids by natural killer T cells. *Nature* **434**, 520–525 (2005)

38. Sriram, V., Du, W., Gervay-Hague, J. & Brutkiewicz, R.R. Cell wall glycosphingolipids of *Sphingomonas paucimobilis* are CD1d-specific ligands for NKT cells. *Eur J Immunol* **35**, 1692–1701 (2005)

39. Neef, A., Witzenberger, R. & Kampfer, P. Detection of sphingomonads and in situ identification in activated sludge using 16S rRNA-targeted oligonucleotide probes. *J Ind Microbiol Biotechnol* **23**, 261–267 (1999)

40. Kawahara, K., Kubota, M., Sato, N., Tsuge, K. & Seto, Y. Occurrence of an alpha-galacturonosyl-ceramide in the dioxin-degrading bacterium *Sphingomonas wittichii*. *FEMS Microbiol Lett* **214**, 289–294 (2002)

41. Kawahara, K., Moll, H., Knirel, Y.A., Seydel, U. & Zahringer, U. Structural analysis of two glycosphingolipids from the lipopolysaccharide-lacking bacterium *Sphingomonas capsulata*. *Eur J Biochem* **267**, 1837–1846 (2000)

42. Hsueh, P.R. Nosocomial infections caused by *Sphingomonas paucimobilis*: clinical features and microbiological characteristics. *Clin Infect Dis* **26**, 676–681 (1998)

43. Perola, O. Recurrent *Sphingomonas paucimobilis*-bacteraemia associated with a multi-bacterial water-borne epidemic among neutropenic patients. *J Hosp Infect* **50**, 196–201 (2002)

44. Kumar, H., Belperron, A., Barthold, S.W. & Bockenstedt, L.K. Cutting edge: CD1d deficiency impairs murine host defense against the spirochete, *Borrelia burgdorferi*. *J Immunol* **165**, 4797–4801 (2000)

45. Ben-Menachem, G., Kubler-Kielb, J., Coxon, B., Yergey, A. & Schneerson, R. A newly discovered cholesteryl galactoside from *Borrelia burgdorferi*. *Proc Natl Acad Sci USA* **100**, 7913–7918 (2003)

46. Kinjo, Y. Natural killer T cells recognize diacylglycerol antigens from pathogenic bacteria. *Nat Immunol* **7**, 978–986 (2006)

47. Michel, M.L. Identification of an IL-17-producing NK1.1(neg) iNKT cell population involved in airway neutrophilia. *J Exp Med* **204**, 995–1001 (2007)

48. Scott-Browne, J.P. Germline-encoded recognition of diverse glycolipids by natural killer T cells. *Nat Immunol* **8**, 1105–1113 (2007)

Ontogeny of the Secondary Antibody Response: Origins and Clonal Diversity

Craig P. Chappell, Joseph Dauner, and Joshy Jacob(✉)

1 Introduction

The adaptive immune response to pathogenic infection or immunization can provide the host with lifelong protection from repeat encounters, a phenomenon known as immunological memory. Immune protection is attributed in large part to the generation of antigen (Ag)-specific T and B lymphocytes during the initial infection that respond with greater rapidity and vigor upon subsequent exposure to the same or crossreactive pathogen.[1] A critical component of this protection is the humoral system, which serves to maintain persistent levels of both circulating serum antibody (Ab) and memory B cells capable of responding to secondary infection. The differentiation of resting memory B cells into antibody-forming cells (AFCs) is widely attributed to the rapid increase in serum Ab levels seen upon secondary infection. Importantly, memory B cell receptors (BCRs) and immune serum often possess an increased affinity for Ag compared to naive B cells, which allows for a qualitatively altered Ab response upon reinfection generally assumed to be more effective at mediating pathogen clearance and/or neutralization. Persistent, circulating Ab therefore provides a first line of defense from invading pathogens, while memory B cells rapidly produce Ab specifically "tailored" for controlling recurrent infections. The quantitative and qualitative improvements found in secondary B cell responses compared to the initial primary response are very effective in mediating protection from disease, as many vaccines in use today rely upon the generation of humoral immunity following immunization.

Following encounter of T-dependent (TD) immunogens (through mechanisms only recently becoming understood), Ag-specific B cells divide and differentiate into either foci of AFCs found throughout the red pulp and follicular borders of the spleen and lymph nodes, or germinal center (GC) B cells located in secondary

Joshy Jacob
Department of Microbiology and Immunology, Emory University, Atlanta, GA 30329, USA

S.P. Schoenberger et al. (eds.) *Crossroads between Innate and Adaptive Immunity II*,
doi: 10.1007/978-0-387-79311-5_4, © Springer Science+Business Media, LLC 2009

follicles.[2] The primary AFC response produces the first wave of Ab following immunization or infection and consists of short-lived, terminally differentiated effector plasma cells that secrete germline-encoded immunoglobulin (Ig) that is often of low affinity.[3,4] Primary AFCs are generated following immunization with either T-dependent or T-independent Ags.

GCs are anatomical structures located within follicular areas of the spleen and lymph nodes where intense B cell proliferation, hypermutation, and Ag-driven selection take place.[3,4] In contrast to primary AFC, these structures are T-dependent and generally arise several days following the appearance of primary AFCs.[2] The conventional model of memory B cell selection within GCs (reviewed in refs.[5-7]) relies upon (1) somatic hypermutation of Ig variable (V) region loci, (2) Ag capture and display by follicular dendritic cells (FDCs), and (3) concurrent positive and negative selection of high- and low-affinity GC B cell clones, respectively. FDCs are myeloid-derived cells specialized for binding and retaining Ag in the form of immune complexes via both $F_{c\gamma}R$-mediated binding of IgG and complement receptors CD21 and CD35.[8-11] FDC-derived survival signals delivered through the BCR and cell surface ligands rescue light zone GC B cells, which otherwise remain prone to negative selection by their continued expression of high amounts of CD95 and c-myc, and low-to-undetectable amounts of Bcl-2 and Bcl-x proteins.[11-19]

Recent live imaging studies in which adoptively transferred Ag-specific B cells were tracked within GCs in vivo revealed that naive B cells frequently enter GCs and survey FDC cell surfaces for Ag. Utilizing Ag-specific B cells of either high or low affinity, it was shown that only B cells with a high affinity for Ag were able participate in ongoing GC reactions, demonstrating the importance of Ag acquisition from FDC for survival within GCs. In addition to FDC-derived signals, cognate interactions with GC CD4 T cells are also critical for memory B cell development. Most notably, surface interactions between CD40–CD40L and ICOS–ICOSL prevent GC B cell apoptosis and contribute to memory B cell selection.[20-23] GC B cells that survive this competitive environment are selected for either a memory B cell or plasma cell fate. AFCs produced in GCs are long-lived cells that reside primarily in the bone marrow and to a lesser extent in secondary lymphoid organs.[24] The combination of somatic hypermutation and selection of high-affinity clones within GCs ultimately results in serum Ab titers and memory B cells consisting of optimized, high-affinity Ab, a process known as affinity maturation.

Germinal center formation and memory B cell development have been extensively studied. Despite the obvious importance that memory B cell activation plays in human health by affording protection from disease, very little effort has been dedicated to anamnestic B cell responses. The main obstacle to studying memory B cells and origins of secondary Ab responses has largely been due to the significant clonal diversity of the initial B cell response itself, as well as an inability to accurately identify and isolate Ag-specific memory B cells. While the conventional model that GC-derived, high-affinity B cells respond to Ag following secondary challenge certainly holds, there are several aspects of secondary Ab responses not adequately explained by this prevailing view. In this mini review, we summarize recent developments in the identification of murine memory B cell subsets, discuss the clonal origins of secondary Ab responses, and suggest a model of secondary Ab responses that incorporates a mechanism for maintaining increased diversity among memory B cells.

2   Development of Memory B Cell Subsets

While the contribution of GCs to the development of memory B cells is undisputed, several reports have noted and characterized the development of Ag-specific memory B cells in the absence of ongoing GC reactions.[19,25,26] Bcl-6-deficient mice are defective in GC formation; however, this did not abolish the formation of NP-binding memory B cells following immunization with NPCG in alum.[25] Sequence analysis of λ_1 L chain V regions from NP-binding memory B cells showed that Bcl-6-deficient mice contained no-to-few mutations in their memory B cell compartment. This exciting observation suggested that memory B cells indeed developed outside GC, but that hypermutation is still restricted to the GC compartment during the course of a typical immune response. These findings were not an artifact of a Bcl-6-deficient environment, as disruption of GCs in C57BL/6 mice through administration of anti-ICOS Ab also yielded a detectable and sizable NP-binding memory B cell compartment that was largely devoid of mutation.[26] Interestingly, treatment with anti-ICOS Ab decreased the size of GCs, but did not affect the somatic hypermutation process within GCs. Despite having affinity-enhancing mutations, mutated GC B cells were not selected into the memory compartment or long-lived bone marrow AFC population, indicating an important role of ICOS-ICOSL interactions and CD4 T cell help for GC-to-memory B cell differentiation. Following adoptive transfer to RAG-deficient hosts, nonmutated memory B cells from anti-ICOS treated mice produced similar numbers of AFC and Ab titers upon Ag challenge compared to B cells from control treated animals. In accord with the lack of mutation, Ab produced by nonmutated memory B cells was of lower affinity than GC-derived memory B cells from control mice. Anti-ICOS treated mice developed persistent ($\geq$day 70), low-affinity serum Ab titers demonstrating that, similar to the memory B cell compartment, low-affinity clones populated the long-lived AFC compartment as well. However, it is not clear from these studies whether BM AFCs were derived from early GC reactions, as anti-ICOS treatment was not initiated until day 6, when GCs have already begun to develop.[2–4,27] Despite the lack of productive GCs in anti-ICOS treated mice, nonmutated memory B cells developed in comparable numbers to control treated mice following primary immunization. The observation that equal numbers of memory B cells developed in ICOS-manipulated mice suggests that entry into the extra-GC developmental pathway is a stochastic process among naive B cells, and not the result of a unique subset of the polyclonal naive B cell repertoire.

Recently, phenotypic analysis of NP-binding memory B cells revealed selectively upregulated on murine memory B cells compared to their naive counterparts. Anderson et al.[28] showed these included both CD80 and CD95 (Fas) on NP-binding B cells.

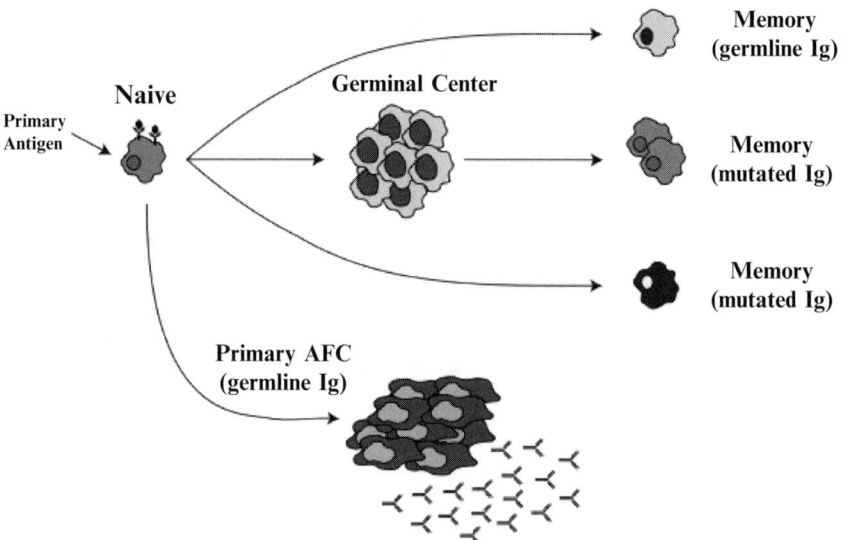

Fig. 1 Development of murine memory B cell subsets. Depicted are memory B cell subsets defined by their hypermutation status and the need for a GC microenvironment to develop. Recent characterizations of both somatic hypermutation and memory B cell selection in the absence of either of GCs and/or immune complexes challenge conventional models of memory B cell development and selection

Furthermore, they were able to segregate CD80$^+$ memory B cells into two subsets based on CD35 expression. While the significance of CD35 expression is not known, both CD80^{neg} and CD80$^+$ CD35$^+$ memory B cells were shown to contain very few, if any, mutations among λ_1 L chains. In contrast, λ_1 L chains from CD80$^+$ CD35^{neg} memory B cells were heavily mutated. These findings both corroborate and further define the nonmutated memory B cell population in the mouse, and lend strong support to the need for amending current models of memory B cell development (Fig. 1).

3 Ongoing Selection of Long-Lived AFC

Following immunization of C57BL/6 mice with NPCG, high-affinity Ag-specific AFCs colonize the BM compartment beginning approximately 14 days postimmunization. These long-lived BM AFCs are derived from GCs.[29] On analysis of V$_H$186.2 gene segments among Ag-binding BM AFCs, they were found to contain approximately 2.5 mutations per V$_H$186.2 gene segment, while GC B cells taken from the same animal contained 5.1 mutations. This suggested that BM AFCs were derived from early GC reactions. Despite the lower frequency of V$_H$ mutation, it was noted that virtually all the BM AFCs contained a canonical affinity-enhancing replacement mutation (Trp ← Leu) at position 33, demonstrating these cells were indeed selected by Ag. These data agree with the notion from several laboratories that AFC differentiation is highly influenced by BCR signal strength.[30,31] Performing similar analyses at day 35 postimmunization revealed that BM AFCs continued to exhibit less mutation within $V_H186.2$ gene segments compared to Ag-specific memory B cells. However, day 35 BM AFCs no longer contained an increased proportion of cells that exhibited the Trp ← Leu mutation at position 33, but instead displayed new mutations indicative of Ag-driven selection. This suggested that BM AFCs are continually recruited during the course of the GC reaction.

It is evident that GCs contain the mutated precursors of both memory B cells in the spleen and high-affinity AFCs in the BM.[32] However, direct seeding of the long-lived BM AFC compartment by B cells emigrating from GCs may not be exclusively responsible, as this does not account for the observations of Takahashi et al.[32] that, following resolution of detectable GCs in the response to NPCG (>day 30 p.i.), the affinity of BM AFCs and serum Ab continued to increase. Additionally, inhibition of GCs by anti-CD40L treatment at early (days 6–10), intermediate (days 10–14), or late (days 16–20) stages of the GC response resulted in the accumulation of increasingly fewer BM AFCs the earlier the administration of treatment, again supporting the notion that early BM AFCs are derived from GCs throughout the course of the GC response. Importantly, despite inhibition of early GCs following treatment and decreased accumulation of total BM AFC numbers, affinity maturation of BM AFCs from day 10 to day 70 was mostly unperturbed. While this continued selection did not require a GC microenvironment, it did require CD4 T cell help.

Additionally, a similar phenomenon was described through VDJ usage and V_H mutation among late-stage (days 35–65) Ag-specific splenic and BM AFCs.[33] In this study, $V_H186.2$ usage and mutation frequencies among splenic AFCs were similar to those seen in GC B cells at day 12 postimmunization. However, while many late-stage Ag-specific BM AFCs were derived from GC B cells (they were V186.2+ with the Trp ← Leu mutation at position 33 and the canonical Try95 mutation), they achieved a much higher frequency of V_H mutations compared to late-stage AFCs from the spleen. Similar to the continued affinity maturation of BM AFCs described earlier, the increase in mutations among BM AFCs was found to depend upon CD4 T cells. Collectively, these studies demonstrate that (1) GC B cells seed the initial BM AFC compartment and, (2) during the immune phase of the response (or following inhibition of GCs), the initial BM AFC clones are increasingly replaced by Ag-selected, high-affinity clones, with the initial BM AFC repertoire becoming largely undetectable by day 60 postimmunization. Given the terminally differentiated nature of AFCs, these findings imply there exists a precursor population of memory B cells that slowly give rise to the long-lived BM AFC compartment during the memory phase of the response. Due to the strong evidence of Ag-driven selection (high mutation frequencies and VDJ usage), this precursor may be a product of GCs. Direct evidence for such a "central memory" population, however, is not very substantial. However, somatic hypermutation in the absence of GCs, as well as Ag-driven selection and memory B cell function without immune complex deposition have been described, although these studies did not analyze BM AFC compartments.[34–37]

4   Diversity of Secondary Ab Responses

The ability to undergo rapid AFC differentiation following secondary Ag encounter is a defining property of memory B cells. Evidence establishing a direct lineage from memory B cells comes from clonal analysis of NP-binding memory B cells and secondary AFCs. These studies revealed that NP-specific memory B cells are dominated by mutated $V_H 186.2^+$ clones and, furthermore, that highly mutated $V_H 186.2^+$ clones appear following secondary responses[38] that indeed resemble those found after the primary response. While it is evident that memory B cells contribute to affinity-matured secondary Ab responses, the extent to which they account for the total secondary Ab response is not clear. Sequence analysis of Ig V_H and V_L segments recovered from AFCs following anti-NP, anti-phOx, and anti-Ars secondary responses has led to findings that demonstrate a *repertoire shift* of Ag-specific Ab following secondary immunization.[39–42] Currently, conventional models of B cell memory do not account for this. For example, concerning the anti-NP secondary response, multiple groups have recovered AFC clones that do not contain canonical $V_H 186.2$–$DFL16.1$–$J_H 2$ rearrangements, but instead utilize analogue V_H gene segments of the $V_H 186.2/V3$ subfamily of V_H genes. In the study described earlier by Lu et al.,[33] mutation frequencies and VDJ usage among splenic AFCs taken during late-stage immune responses (days 35–65) were compared to those obtained from AFCs following secondary challenges (day 4 postsecondary). They found analogue clones accounted for ~20% of secondary AFCs following Ag rechallenge. Furthermore, earlier studies analyzing hybridomas generated from secondary AFCs showed that while the majority of Ag-specific hybridomas utilized $V_H 186.2$ gene segments rearranged to DFL16.1, many of these clones contained CDR3 rearrangements with more extensive nucleotide deletions at the D–J boundary than those described in studies of GC B cells.[4,43–45] Furthermore, a similar finding by Siekevitz et al.[46] using hybridomas derived from secondary anti-NP AFCs revealed a pattern of nucleotide deletions within D segments similar to the clones described in the studies above. In addition to canonical IgH chain usage among anti-NP memory B cells in mice of the *Ighb* haplotype, >98% of anti-NP serum Ab 40days following immunization utilize λ_1 L chains (Fig. 2a). In response to secondary NPCG immunization, the anti-NP λ_1 light chain response increases ~6.5-fold, whereas anti-NP utilizing κ L chains increases 34-fold. Since memory B cells utilizing κ L chains are virtually undetectable among late-stage GCs, AFCs, or memory B cells, the origin of these cells remains unclear. One possibility is that conventional methods of detecting Ag-specific memory B cells and AFC allow a significant portion of the immune B cell repertoire to go undetected. With the exception of λ_1 expression in the NP system, most techniques used to isolate individual clones of hapten-specific memory B cells or AFCs rely upon Ag binding, either directly (flow cytometry) or through Abs produced by AFCs (hemolytic plaque assay). While these assays have allowed consistent detection of high-affinity Ag-specific memory B cells and AFCs, the possibility that some Ag-specific clones will lack sufficient affinity to facilitate detection will always remain open.

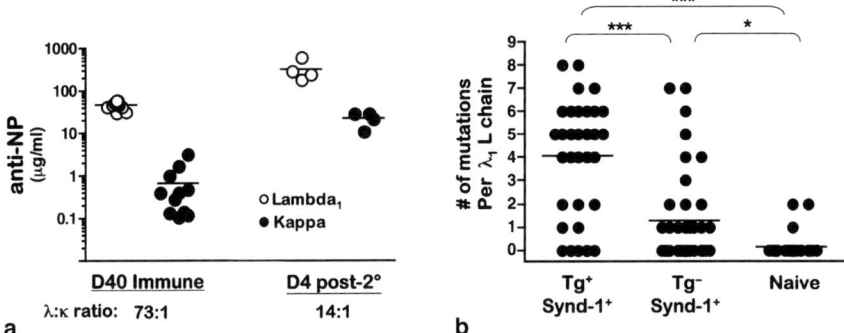

Fig. 2 Ab repertoire shift following secondary NPCG challenge in C57BL/6 mice. (**a**) Cohorts of C57BL/6 mice were immunized i.p. with 50-μg alum-precipitated NPCG and allowed to rest for 40days. Serum samples were taken and four mice were rechallenged i.v. with 20-μg NPCG in PBS. Four days following rechallenge serum samples were drawn again. The quantity of anti-NP serum Ab utilizing either λ_1 or κL chains is plotted. (**b**) Four days postsecondary NPCG challenge spleens were harvested and sorted into Synd-1⁺ Tg⁺ and Synd-1⁺ Tg⁻ populations. λ_1 sequences were PCR amplified, cloned, and individual colonies chosen for sequence analysis. Scatter plot shows the number of unique mutations within each sequenced clone. Transgene (Tg) refers to either β-gal or YFP expression. *$p < 0.05$ and ***$p < 0.001$ by Kruskal–Wallis test

Our laboratory has recently generated a novel mouse model (GCCxR26R) to induce permanent transgene (Tg) expression solely in GC B cells.[48] This strategy allowed us to identify the long-lived progeny of GC B cells without relying upon Ag-labeling techniques, giving the unique advantage of identifying Ag-specific, GC-derived memory B cells without regard to Ag receptor affinity. While Ag rechallenge experiments demonstrated the expected expansion and differentiation of β-gal⁺ memory B cells, we were surprised to find that 70–75% of the total secondary AFC response consisted of non-GC-derived (i.e., β-gal⁻) AFCs. Interestingly, sequence analysis of λ_1 L chains from sorted β-gal⁺ and β-gal⁻ AFCs revealed that a significant proportion (21/40) of β-gal⁻ secondary AFCs were nonmutated (Fig. 2b). In contrast, only 5/34 β-gal⁺ AFC clones lacked mutations, with most cells displaying between four and eight mutations. Calculating to obtain absolute numbers of nonmutated, λ_1^+ anti-NP AFCs per spleen, we estimate that almost half (~44%) of the secondary anti-NP AFCs express germline Ab.

Unexpectedly, the GCCxR26R memory B cell mouse model did not label all GC B cells by conventional phenotypic marker analysis; therefore, the possibility that nonmutated, GC-derived memory B cells gave rise to these cells remains open. However, it is considerably unlikely that a significant proportion of nonmutated B cells are undergoing positive selection in GCs. Considering the number of recent reports characterizing nonmutated memory B cell development outside GCs discussed earlier, a portion of this response is likely derived from these cells. However, estimates of the size of the nonmutated memory B cell population following NPCG immunization are placed at 15–20% of the total NP-binding memory pool.[19,26,28]

Considering our finding that >40% of total secondary anti-NP AFCs utilizing λ_1 L chains in GCCxR26R mice contained germline V regions, we feel it is unlikely that non-GC-derived memory B cells account for the entire germline secondary Ab response. Additionally, it is important to note that NP-binding memory B cells are almost exclusively λ_1^+. We have recently found that ~40% of β-gal⁻ secondary AFCs utilized κ L chains to form anti-NP Ab, demonstrating that a significant portion the secondary Ab response derives from precursor cells that are not detected by conventional Ag-binding techniques (Chappell et al., manuscript in preparation). Experiments are currently underway to determine the relative affinity of κ^+ vs. λ_1^+ anti-NP Ab secreted following secondary Ag challenge with NPCG.

5   Immune Complex-Mediated Enhancement of Ab Responses

Antibodies are the effector molecules of B cells. While the capacity to neutralize pathogens through direct binding is generally the mechanism one thinks of when considering Ab effector function, this is only one manner in which Abs mediate their effectiveness. Additional effector functions are largely attributed to activation of the complement system and isotype-specific F_c receptor-mediated binding to effector cells. When complexed with specific Ag (i.e., immune complexes (ICs)), Abs facilitate a diverse array of functions including Ag clearance, NK cell activation, Ag transport across mucosal barriers, exacerbation of allergic reactions and autoimmunity, and regulation of B cell responses to Ag. It has been known for some time that passive transfer of antiserum prior to immunization has a profound capacity to alter the ensuing immune response. In some circumstances, enhancement of the Ag-specific serum Ab response is observed,[49–51] while other regimens produced a suppressive effect.[52,53] The isotype and Ag:Ab ratio requirements for this phenomenon were investigated during the 1970s and 1980s when it became evident that immunization with immune complexes formed in vitro produced robust priming for GC and memory B cell responses.[54–63] In contrast, several studies of that period demonstrated Ag-specific suppression of B cell responses following administration of Ag:Ab complexes.[52,53] Since these early studies differed from one another with respect to Ab source, Ag choice, IC preparation protocols, immunization route, etc., it was not until the wide availability of monoclonal Abs and genetically manipulated mice that these pathways were dissected.

First, it was revealed that inhibition vs. activation of the response depended upon both the isotype of Ab used to form the ICs, as well as whether the Ag was in soluble or particulate form. For example, when Ag-specific IgM was used to make ICs, it was found to enhance Ab responses only when combined with particulate Ag.[64] On the other hand, when IgG was used form ICs, the reverse situation was observed. When in complex with particulate sheep erythrocytes (SRBC), IgG Ab mediates suppression of Ab responses.[53,65] This effect is not specific to SRBC, as suppression is also seen when administering Ag in adjuvants.[66–69] In contrast, enhancement of Ab responses and numbers of Ag-specific AFC are seen when IgG is in complex with soluble Ags, such as TNP–BSA or TNP–KLH.[63,70–73] The reasons governing activation vs. suppression are not entirely clear (see later). Nevertheless, concerning the enhancement of responses, it was subsequently demonstrated that enhancement with IgM-containing ICs requires complement receptors Cr1/2, while IgG-containing ICs require $F_{C}\gamma R$ expression.[70,74,75] Specific F_{C} receptors exist for IgG and IgE isotypes and are widely expressed in the body, particularly among cells of the hematopoietic system. F_{C} receptor expression is found on mast cells, B cells, DCs, FDCs, and macrophages. B cells constitutively express $F_{C\gamma}RIIB$, which contains cytoplasmic ITIM signaling motifs capable of mediating negative signaling events (reviewed in ref. [76]). However, the remaining F_{C} receptors ($F_{C\gamma}RI$, $F_{C\gamma}RIII$, $F_{C}\varepsilon RI$, $F_{C}\varepsilon RII$, and $F_{C}\varepsilon RIII$) possess costimulation and phagocytic properties and are not involved in downregulation of the immune response. Complement receptors CD21 and CD35 are also found on B cells and FDCs. Expression of CD21 is critical for optimal GC B cell responses, as *Cr1/2*-deficient mice do not raise efficient GC or recall responses.[77,78]

It is almost certain that long-term Ag retention requires $F_{C}R$ and Cr1/2 expression, and is most often postulated to occur on the surface of FDCs given the fact their high expression of $F_{C\gamma}RI$, $F_{C\gamma}RIII$, CR1, and CR2 receptors. Despite the proposed role for long-term FDC-mediated retention of Ag, FDC-associated $F_{C\gamma}R$ expression was surprisingly *not* required for IC-mediated enhancement of Ab responses.[70] Elegant bone marrow chimera experiments using $F_{C\gamma}R$-deficient mice showed that $F_{C\gamma}R$ expression was required on cells of lymphoid origin (FDC are myeloid).[79] While the precise cell type responsible for binding and presenting ICs is not known, a likely candidate are lymphoid-derived DCs, which also express $F_{C\gamma}Rs$.

In addition to enhancement of primary Ab responses, immunization with ICs was shown to produce more numerous GCs than immunization with Ag alone. Ag:Ab complexes of correct Ig isotype are potent activators of the classical complement pathway. Given the abundance of data demonstrating that CD21 (CR2) expression together with CD19 and CD81 in the BCR coreceptor complex lowers the signaling threshold of BCR-mediated activation, it may not be surprising that the diversity of B cell clones participating in GCs was greater in mice that were immunized with NP-containing ICs compared to those immunized with NPCG in alum.[80,81] Since B cells do not express any of the activating $F_{C\gamma}Rs$, this result implies involvement of the complement components. However, it should be noted ICs were formed with anti-NP IgG1 mAb, which do not have a high capacity to activate the classical complement pathway. Whether this reflects activation of the alternative complement pathway (as the authors proposed), or the fact that $F_{C\gamma}Rs$ are sufficient to mediate the effects seen, is not known. Finally, we could find no reports that determine whether the diversity of GC B cells following IC immunization is also reflected in the memory B cell repertoire or secondary AFC response.

Based on the ability of ICs to enhance B cell activation and differentiation when administered in vivo, it seems highly plausible that repertoire shifts occurring during secondary Ab responses may be derived from naive B cells. This notion is supported by using an anti-idiotype Ab as immunogen to induce secondary responses in NP-primed mice. Siekevitz et al.[46] detected the rapid appearance of

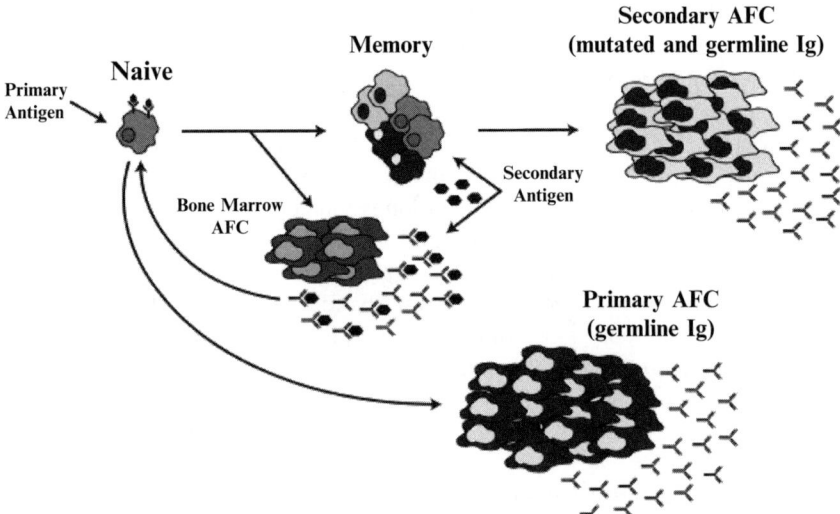

Fig. 3 Immune complex activation of naive B cells during secondary Ab responses. Schematic diagram showing a proposed model for the origins of secondary Ab repertoire shifts. The formation of ICs upon secondary immunization enhances naive B cell activation and differentiation. The mechanism(s) by which this occurs in vivo is poorly understood, but likely reflects, in part, increased CD4 T cell activation via rapid processing of Ag by $F_{c\gamma}R^+$ lymphoid DCs

AFC clones utilizing novel VDJ rearrangements in the secondary response. Since these clones were not detected following secondary challenge with the native Ag, the authors attributed their appearance to the activation of naive B cells (Fig. 3). The precise mechanism by which this occurred was not investigated, but nevertheless demonstrates the recruitment of naive B cells following a secondary challenge. Together, these results indicate either rapid recruitment of naïve B cell clones during secondary responses, or activation of memory clones that have previously gone undetected in the primary response.

In addition to rapid AFC differentiation, GCs also appear following secondary immunization. Although their rapid appearance would suggest derivation from memory B cells, there is conflicting evidence in the literature concerning whether memory B cells can seed secondary GCs. Studies analyzing anti-phOx responses demonstrated that Ag-specific GC B cells isolated following secondary immunization contained progressive increases in mutation frequencies.[82] However, this finding seems to stand alone, as other attempts to analyze secondary GC responses have yielded results that do not implicate participation by memory B cells. This is supported by experiments showing no additional mutations accrued in the Ag-specific memory B cell pool following secondary Ag challenge of adoptively transferred NPCG-immune splenocytes. This led the authors to conclude that memory B cells are resistant to Ig hypermutation.[46] Furthermore, in a study assessing the requirement of CD4 T cell help for memory B cell activation, Hebeis et al.[83]

showed that, after Ag challenge of recipients of memory B cells (and CD4 help), the resulting GCs followed primary response kinetics and consisted of host-derived B cells, not B cells transferred from immune animals. The transferred memory B cells were present and capable of responding, however, as they gave rise to a rapid AFC response following Ag challenge. These latter studies argue against a role for memory B cell participation in secondary GC formation and agree with the notion that further somatic hypermutation events are unlikely to improve upon the affinity of a B cell that has already undergone affinity maturation and selection.

6 Concluding Remarks

In this review, we have attempted to highlight studies, both recent and not so recent, which support the notion that conventional models of memory B cell development and secondary Ab responses are in need of amendment. In addition to the clonal origins of secondary Ab responses, many questions surrounding the differentiation products of memory B cells remain unanswered. The ability to re-enter GCs, self-renew, or differentiate to preplasma cell intermediates following activation by secondary Ag are all areas that need further investigation. Recent developments in genetic mouse models to detect Ag-specific memory B cells, coupled with gene array technology, promise to provide a wealth of information in the future concerning memory B cell development, selection, activation, and differentiation.

References

1. Gray, D. Immunological memory. *Annu Rev Immunol* **11,** 49–77 (1993)
2. Jacob, J., Kassir, R. & Kelsoe, G. In situ studies of the primary immune response to (4-hydroxy-3-nitrophenyl)acetyl. I. The architecture and dynamics of responding cell populations. *J Exp Med* **173,** 1165–1175 (1991)
3. Jacob, J. & Kelsoe, G. In situ studies of the primary immune response to (4-hydroxy-3-nitrophenyl)acetyl. II. A common clonal origin for periarteriolar lymphoid sheath-associated foci and germinal centers. *J Exp Med* **176,** 679–687 (1992)
4. Jacob, J., Kelsoe, G., Rajewsky, K. & Weiss, U. Intraclonal generation of antibody mutants in germinal centres. *Nature* **354,** 389–392 (1991)
5. Cozine, C.L., Wolniak, K.L. & Waldschmidt, T.J. The primary germinal center response in mice. *Curr Opin Immunol* **17,** 298–302 (2005)
6. Wolniak, K.L., Shinall, S.M. & Waldschmidt, T.J. The germinal center response. *Crit Rev Immunol* **24,** 39–65 (2004)
7. McHeyzer-Williams, L.J. & McHeyzer-Williams, M.G. Antigen-specific memory B cell development. *Annu Rev Immunol* **23,** 487–513 (2005)
8. Fang, Y., Xu, C., Fu, Y.X., Holers, V.M. & Molina, H. Expression of complement receptors 1 and 2 on follicular dendritic cells is necessary for the generation of a strong antigen-specific IgG response. *J Immunol* **160,** 5273–5279 (1998)
9. Humphrey, J.H., Grennan, D. & Sundaram, V. The origin of follicular dendritic cells in the mouse and the mechanism of trapping of immune complexes on them. *Eur J Immunol* **14,** 859–864 (1984)

10. Qin, D. et al. Evidence for an important interaction between a complement-derived CD21 ligand on follicular dendritic cells and CD21 on B cells in the initiation of IgG responses. *J Immunol* **161,** 4549–4554 (1998)

11. Tew, J.G., Wu, J., Fakher, M., Szakal, A.K. & Qin, D. Follicular dendritic cells: beyond the necessity of T-cell help. *Trends Immunol* **22,** 361–367 (2001)

12. Cutrona, G. et al. The propensity to apoptosis of centrocytes and centroblasts correlates with elevated levels of intracellular myc protein. *Eur J Immunol* **27,** 234–238 (1997)

13. Inada, K. et al. c-Fos induces apoptosis in germinal center B cells. *J Immunol* 161, 3853–3861 (1998)

14. Liu, Y.J. et al. Germinal center cells express bcl-2 protein after activation by signals which prevent their entry into apoptosis. *Eur J Immunol* **21,** 1905–1910 (1991)

15. Martinez-Valdez, H. et al. Human germinal center B cells express the apoptosis-inducing genes Fas, c-myc, P53, and Bax but not the survival gene bcl-2. *J Exp Med* **183,** 971–977 (1996)

16. Shokat, K.M. & Goodnow, C.C. Antigen-induced B-cell death and elimination during germinal-centre immune responses. *Nature* **375,** 334–338 (1995)

17. Tuscano, J.M. et al. Bcl-*x* rather than Bcl-2 mediates CD40-dependent centrocyte survival in the germinal center. *Blood* **88,** 1359–1364 (1996)

18. Wang, Z., Karras, J.G., Howard, R.G. & Rothstein, T.L. Induction of bcl-*x* by CD40 engagement rescues sIg-induced apoptosis in murine B cells. *J Immunol* **155,** 3722–3725 (1995)

19. Takahashi, Y., Ohta, H. & Takemori, T. Fas is required for clonal selection in germinal centers and the subsequent establishment of the memory B cell repertoire. *Immunity* **14,** 181–192 (2001)

20. Lebecque, S., de Bouteiller, O., Arpin, C., Banchereau, J. & Liu, Y.J. Germinal center founder cells display propensity for apoptosis before onset of somatic mutation. *J Exp Med* **185,** 563–571 (1997)

21. Cleary, A.M., Fortune, S.M., Yellin, M.J., Chess, L. & Lederman, S. Opposing roles of CD95 (Fas/APO-1) and CD40 in the death and rescue of human low density tonsillar B cells. *J Immunol* **155,** 3329–3337 (1995)

22. Laman, J.D., Claassen, E. & Noelle, R.J. Functions of CD40 and its ligand, gp39 (CD40L). *Crit Rev Immunol* **16,** 59–108 (1996)

23. Choe, J., Kim, H.S., Zhang, X., Armitage, R.J. & Choi, Y.S. Cellular and molecular factors that regulate the differentiation and apoptosis of germinal center B cells. Anti-Ig down-regulates Fas expression of CD40 ligand-stimulated germinal center B cells and inhibits Fas-mediated apoptosis. *J Immunol* **157,** 1006–1016 (1996)

24. Slifka, M.K., Antia, R., Whitmire, J.K. & Ahmed, R. Humoral immunity due to long-lived plasma cells. *Immunity* **8,** 363–372 (1998)

25. Toyama, H. et al. Memory B cells without somatic hypermutation are generated from Bcl6-deficient B cells. *Immunity* **17,** 329–339 (2002)

26. Inamine, A. et al. Two waves of memory B-cell generation in the primary immune response. *Int Immunol* **17,** 581–589 (2005)

27. McHeyzer-Williams, M.G., McLean, M.J., Lalor, P.A. & Nossal, G.J. Antigen-driven B cell differentiation in vivo. *J Exp Med* **178,** 295–307 (1993)

28. Anderson, S.M., Tomayko, M.M., Ahuja, A., Haberman, A.M. & Shlomchik, M.J. New markers for murine memory B cells that define mutated and unmutated subsets. *J Exp Med* **204,** 2103–2114 (2007)

29. Smith, K.G., Light, A., Nossal, G.J. & Tarlinton, D.M. The extent of affinity maturation differs between the memory and antibody-forming cell compartments in the primary immune response. *EMBO J* **16,** 2996–3006 (1997)

30. Paus, D. et al. Antigen recognition strength regulates the choice between extrafollicular plasma cell and germinal center B cell differentiation. *J Exp Med* **203,** 1081–1091 (2006)

31. Shih, T.A., Meffre, E., Roederer, M. & Nussenzweig, M.C. Role of BCR affinity in T cell dependent antibody responses in vivo. *Nat Immunol* **3,** 570–575 (2002)

32. Takahashi, Y., Dutta, P.R., Cerasoli, D.M. & Kelsoe, G. In situ studies of the primary immune response to (4-hydroxy-3-nitrophenyl)acetyl. V. Affinity maturation develops in two stages of clonal selection. *J Exp Med* **187,** 885–895 (1998)

33. Lu, Y.F., Singh, M. & Cerny, J. Canonical germinal center B cells may not dominate the memory response to antigenic challenge. *Int Immunol* **13**, 643–655 (2001)
34. William, J., Euler, C., Christensen, S. & Shlomchik, M.J. Evolution of autoantibody responses via somatic hypermutation outside of germinal centers. *Science* **297**, 2066–2070 (2002)
35. Rossbacher, J., Haberman, A.M., Neschen, S., Khalil, A. & Shlomchik, M.J. Antibody-independent B cell-intrinsic and -extrinsic roles for CD21/35. *Eur J Immunol* **36**, 2384–2393 (2006)
36. Anderson, S.M., Hannum, L.G. & Shlomchik, M.J. Memory B cell survival and function in the absence of secreted antibody and immune complexes on follicular dendritic cells. *J Immunol* **176**, 4515–4519 (2006)
37. Hannum, L.G., Haberman, A.M., Anderson, S.M. & Shlomchik, M.J. Germinal center initiation, variable gene region hypermutation, and mutant B cell selection without detectable immune complexes on follicular dendritic cells. *J Exp Med* **192**, 931–942 (2000)
38. Weiss, U. & Rajewsky, K. The repertoire of somatic antibody mutants accumulating in the memory compartment after primary immunization is restricted through affinity maturation and mirrors that expressed in the secondary response. *J Exp Med* **172**, 1681–1689 (1990)
39. Berek, C., Griffiths, G.M. & Milstein, C. Molecular events during maturation of the immune response to oxazolone. *Nature* **316**, 412–418 (1985)
40. Berek, C. & Milstein, C. Mutation drift and repertoire shift in the maturation of the immune response. *Immunol Rev* **96**, 23–41 (1987)
41. Griffiths, G.M., Berek, C., Kaartinen, M. & Milstein, C. Somatic mutation and the maturation of immune response to 2-phenyl oxazolone. *Nature* **312**, 271–275 (1984)
42. Blier, P.R. & Bothwell, A.L. The immune response to the hapten NP in C57BL/6 mice: insights into the structure of the B-cell repertoire. *Immunol Rev* **105**, 27–43 (1988)
43. Blier, P.R. & Bothwell, A. A limited number of B cell lineages generates the heterogeneity of a secondary immune response. *J Immunol* **139**, 3996–4006 (1987)
44. Cumano, A. & Rajewsky, K. Structure of primary anti-(4-hydroxy-3-nitrophenyl)acetyl (NP) antibodies in normal and idiotypically suppressed C57BL/6 mice. *Eur J Immunol* **15**, 512–520 (1985)
45. Jacob, J., Przylepa, J., Miller, C. & Kelsoe, G. In situ studies of the primary immune response to (4-hydroxy-3-nitrophenyl)acetyl. III. The kinetics of V region mutation and selection in germinal center B cells. *J Exp Med* **178**, 1293–1307 (1993)
46. Siekevitz, M., Kocks, C., Rajewsky, K. & Dildrop, R. Analysis of somatic mutation and class switching in naive and memory B cells generating adoptive primary and secondary responses. *Cell* **48**, 757–770 (1987)
47. Smith, F.I., Cumano, A., Licht, A., Pecht, I. & Rajewsky, K. Low affinity of kappa chain bearing (4-hydroxy-3-nitrophenyl)acetyl (NP)-specific antibodies in the primary antibody repertoire of C57BL/6 mice may explain lambda chain dominance in primary anti-NP responses. *Mol Immunol* **22**, 1209–1216 (1985)
48. Chappell, C.P. & Jacob, J. Identification of memory B cells using a novel transgenic mouse model. *J Immunol* **176**, 4706–4715 (2006)
49. Terres, G. & Wolins, W. Enhanced sensitization in mice by simultaneous injection of antigen and specific rabbit antiserum. *Proc Soc Exp Biol Med* **102**, 632–635 (1959)
50. Terres, G. & Wolins, W. Enhanced immunological sensitization of mice by the simultaneous injection of antigen and specific antiserum. I. Effect of varying the amount of antigen used relative to the antiserum. *J Immunol* **86**, 361–368 (1961)
51. Terres, G. & Stoner, R.D. Specificity of enhanced immunological sensitization of mice following injections of antigens and specific antisera. *Proc Soc Exp Biol Med* **109**, 88–91 (1962)
52. Collisson, E.W., Andersson, B. & Lamon, E.W. Modulation of hapten-specific antibody responses with anticarrier antibody. I. Differential effects of IgM and IgG anticarrier on primary direct and indirect hapten-specific plaque-forming cells. *Proc Soc Exp Biol Med* **162**, 194–198 (1979)
53. Heyman, B. & Wigzell, H. Specific IgM enhances and IgG inhibits the induction of immunological memory in mice. *Scand J Immunol* **21**, 255–266 (1985)

54. Laissue, J., Cottier, H., Hess, M.W. & Stoner, R.D. Early and enhanced germinal center formation and antibody responses in mice after primary stimulation with antigen–isologous antibody complexes as compared with antigen alone. *J Immunol* **107,** 822–831 (1971)

55. Klaus, G.G. The generation of memory cells. II. Generation of B memory cells with preformed antigen–antibody complexes. *Immunology* **34,** 643–652 (1978)

56. Klaus, G.G. Generation of memory cells. III. Antibody class requirements for the generation of B-memory cells by antigen–antibody complexes. *Immunology* **37,** 345–351 (1979)

57. Klaus, G.G. & Humphrey, J.H. The generation of memory cells. I. The role of C3 in the generation of B memory cells. *Immunology* **33,** 31–40 (1977)

58. Kunkl, A. & Klaus, G.G. The generation of memory cells. IV. Immunization with antigen-antibody complexes accelerates the development of B-memory cells, the formation of germinal centres and the maturation of antibody affinity in the secondary response. *Immunology* **43,** 371–378 (1981)

59. Kunkl, A. & Klaus, G.G. The generation of memory cells. V. Preferential priming of IgG1 B memory cells by immunization with antigen IgG2 antibody complexes. *Immunology* **44,** 163–168 (1981)

60. Stoner, R.D., Terres, G. & Hess, M.W. Early and enhanced antioxin responses elicited with complexes of tetanus toxoid and specific mouse and human antibodies. *J Infect Dis* **131,** 230–238 (1975)

61. Terres, G., Habicht, G.S. & Stoner, R.D. Carrier-specific enhancement of the immune response using antigen–antibody complexes. *J Immunol* **112,** 804–811 (1974)

62. Terres, G., Morrison, S.L., Habicht, G.S. & Stoner, R.D. Appearance of an early "primed state" in mice following the concomitant injections of antigen and specific antiserum. *J Immunol* **108,** 1473–1481 (1972)

63. Coulie, P.G. & Van Snick, J. Enhancement of IgG anti-carrier responses by IgG2 anti-hapten antibodies in mice. *Eur J Immunol* **15,** 793–798 (1985)

64. Henry, C. & Jerne, N.K. Competition of 19S and 7S antigen receptors in the regulation of the primary immune response. *J Exp Med* **128,** 133–152 (1968)

65. Quintana, I.Z., Silveira, A.V. & Moller, G. Regulation of the antibody response to sheep erythrocytes by monoclonal IgG antibodies. *Eur J Immunol* **17,** 1343–1349 (1987)

66. Cerottini, J.C., McConahey, P.J. & Dixon, F.J. Specificity of the immunosuppression caused by passive administration of antibody. *J Immunol* **103,** 268–275 (1969)

67. Karlsson, M.C., Wernersson, S., Diaz de Stahl, T., Gustavsson, S. & Heyman, B. Efficient IgG-mediated suppression of primary antibody responses in Fcgamma receptor-deficient mice. *Proc Natl Acad Sci USA* **96,** 2244–2249 (1999)

68. Krieger, N.J., Pesce, A. & Michael, J.G. Immunoregulation of the anti-bovine serum albumin response by polyclonal and monoclonal antibodies. *Cell Immunol* **80,** 279–287 (1983)

69. Strannegard, O. & Belin, L. Suppression of reagin synthesis in rabbits by passively administered antibody. *Immunology* **18,** 775–785 (1970)

70. Wernersson, S. IgG-mediated enhancement of antibody responses is low in Fc receptor gamma chain-deficient mice and increased in Fc gamma RII-deficient mice. *J Immunol* **163,** 618–622 (1999)

71. Wiersma, E.J. Enhancement of the antibody response to protein antigens by specific IgG under different experimental conditions. *Scand J Immunol* **36,** 193–200 (1992)

72. Wiersma, E.J., Nose, M. & Heyman, B. Evidence of IgG-mediated enhancement of the antibody response in vivo without complement activation via the classical pathway. *Eur J Immunol* **20,** 2585–2589 (1990)

73. Enriquez-Rincon, F. & Klaus, G.G. Differing effects of monoclonal anti-hapten antibodies on humoral responses to soluble or particulate antigens. *Immunology* **52,** 129–136 (1984)

74. Applequist, S.E., Dahlstrom, J., Jiang, N., Molina, H. & Heyman, B. Antibody production in mice deficient for complement receptors 1 and 2 can be induced by IgG/Ag and IgE/Ag, but not IgM/Ag complexes. *J Immunol* **165,** 2398–2403 (2000)

75. Heyman, B., Pilstrom, L. & Shulman, M.J. Complement activation is required for IgM-mediated enhancement of the antibody response. *J Exp Med* **167,** 1999–2004 (1988)

76. Ravetch, J.V. & Kinet, J.P. Fc receptors. *Annu Rev Immunol* **9,** 457–492 (1991)
77. Ahearn, J.M. et al. Disruption of the Cr2 locus results in a reduction in B-1a cells and in an impaired B cell response to T-dependent antigen. *Immunity* **4,** 251–262 (1996)
78. Fischer, M.B. et al. Dependence of germinal center B cells on expression of CD21/CD35 for survival. *Science* **280,** 582–585 (1998)
79. Diaz de Stahl, T. & Heyman, B. IgG2a-mediated enhancement of antibody responses is dependent on FcRgamma+ bone marrow-derived cells. *Scand J Immunol* **54,** 495–500 (2001)
80. Nie, X., Basu, S. & Cerny, J. Immunization with immune complex alters the repertoire of antigen-reactive B cells in the germinal centers. *Eur J Immunol* **27,** 3517–3525 (1997)
81. Song, H., Nie, X., Basu, S., Singh, M. & Cerny, J. Regulation of VH gene repertoire and somatic mutation in germinal centre B cells by passively administered antibody. *Immunology* **98,** 258–266 (1999)
82. Rada, C., Gupta, S.K., Gherardi, E. & Milstein, C. Mutation and selection during the secondary response to 2-phenyl oxazolone. *Proc Natl Acad Sci USA* **88,** 5508–5512 (1991)
83. Hebeis, B.J. et al. Activation of virus-specific memory B cells in the absence of T cell help. *J Exp Med* **199,** 593–602 (2004)

The Role of PI3K Signalling in the B Cell Response to Antigen

Daniel J. Hodson and Martin Turner(⊠)

1   Introduction

The PI3K signal transduction pathway is employed by a wide variety of cells in organisms from yeasts to humans to regulate biological processes such as growth, metabolism and survival. In B cells the PI3K signalling pathway is activated downstream of the B cell receptor where, in contrast to cells outside the immune system, signalling appears particularly dependent upon the PI3K catalytic subunit known as p110δ. BCR-induced PI3K activity is believed to play an important role in the regulation of both B cell development and the B cell response to antigen. Although we have learned much from the use of gene-targeted mice several areas of uncertainty still exist. Our current knowledge of the role of PI3K signalling in the B cell antigen response will be discussed.

2   Background to the PI3K Signal Transduction Pathway

The phosphoinositide-3-kinases (PI3Ks) are a highly conserved family of lipid kinases that phosphorylate the 3′-hydroxyl group of phosphoinositide membrane lipids. They are activated by the effects of extracellular stimuli on cell surface receptors, and ultimately lead to the recruitment of intracellular proteins that regulate diverse biological processes but typically promote cell growth, proliferation, survival and differentiation. They are employed by an array of receptors (including growth factor, antigen and cytokine receptors) in a wide variety of cell types from yeasts through to humans. Although PI3Ks are divided into several classes most information exists about class I.

Martin Turner
Laboratory of Lymphocyte Signalling and Development, Babraham Institute,
Cambridge, UK

S.P. Schoenberger et al. (eds.) *Crossroads between Innate and Adaptive Immunity II*,
doi: 10.1007/978-0-387-79311-5_5, © Springer Science+Business Media, LLC 2009

Class Ia PI3Ks are activated downstream of receptor-associated tyrosine kinases such as syk or ZAP-70. They exist as heterodimers of one catalytic subunit (p110α, p110β or p110δ) and one (or occasionally two) regulatory subunit (p85αa, p55α, p50α, p85β or p55γ). There are important differences between the three catalytic isoforms (discussed later); however, the five regulatory subunits, three of which are splice variants of the same gene, are considered to be functionally equivalent and are collectively known as 'p85s'. The regulatory subunits contain src homology-2 (SH2) domains which allow binding to phosphorylated tyrosine residues within YXXM motifs found in cytoplasmic tails of the cell surface receptor complexes. Class Ib PI3Ks are activated downstream of G protein-coupled receptors and consist of a catalytic p110γ and a regulatory subunit (p101 or p84). All class I PI3Ks catalyse the phosphorylation of phosphatidylinositol-(4,5)-biphosphate ($PI(4,5)P_2$) to produce the active second messenger phosphatidylinositol-(3,4,5)-triphosphate ($PI(3,4,5)P_3$).[14] For simplicity these two molecules are often referred to as PIP_2 and PIP_3, respectively. Downstream effector proteins that possess a pleckstrin homology (PH) domain are then able to bind to PIP_3 and are recruited to the signalling complex. One of the most important PH domain-containing proteins recruited by PIP_3 is protein kinase B (PKB – also known as AKT).[1] PKB is responsible for many of the effects of PI3K signalling in a wide range of cell types. The activity of PI3K is opposed by the constitutively active lipid phosphatase PTEN and the inducible lipid phosphatase SHIP.[17] PTEN catalyses the hydrolysis of $PI(3,4,5)P_3$ to $PI(4,5)P_2$ while SHIP catalyses the hydrolysis of $PI(3,4,5)P_3$ to $PI(3,4)P_2$ which may itself contribute further to PKB activation or the function of other PH domain-containing proteins.[28] Mutation or deletion of the PTEN gene results in constitutive activation of the PI3K pathway and is one of the most common tumour suppressor mutations seen in human cancer.[31]

3   PI3K Signalling in B Cells

In B lymphocytes PI3K is activated downstream of the BCR, CD40, TLR and cytokine receptors. The BCR is a particularly strong activator of PI3K activity. BCR engagement leads to the activation of the tyrosine kinase syk which mediates phosphorylation of tyrosine within YXXM motifs.[12] Although not present in the BCR itself these YXXM motifs are found in the cytoplasmic tails of other components of the BCR signalling complex including the complement receptor CD19 and B cell adaptor for phosphoinositide-3-kinase (BCAP).[20,37] Phosphorylated YXXM motifs bind the SH2 domains of the p85 regulatory subunits. Thus PI3K is recruited into the BCR signalling complex by BCAP and CD19. Consistent with this B cell PI3K activation is greatly increased by co-engagement of the BCR and the CD19 complement receptor rendering B cells much more sensitive to antigen encountered in the presence of complement.[9] We have shown P110δ is crucial for the synergistic signal integration between CD19 and membrane immunoglobulin for activation of PKB and ERK.[35,36] The recruitment of PI3K to the cell membrane from its resting location in the cytoplasm brings it into close approximation with its substrate PIP_2 which it phosphorylates to generate the active second messenger PIP_3. BCR ligation leads to rapid elevation of PIP_3 levels peaking within 2 min, then reverting to baseline within 10 min.[7] PIP_3 binds to PH domains and thus serves as a membrane-docking platform to recruit a number of PH domain-containing effector proteins. Most well characterised of these is PKB/AKT, but others include PDK, BTK and various guanine nucleotide exchange factors. PDK1 itself phosphorylates and activates PKB.[1] Phosphorylated PKB is a useful marker of cellular PI3K activation status. Downstream effects of PKB activation include repression of the transcription factor FOXO, activation of mTOR and the NF-κB pathway.[14] As in other cells the PI3K system is opposed by the phosphatase activity of PTEN and SHIP 1 and 2. Levels of PIP_3 and phosphorylated PKB are elevated in both resting and BCR-stimulated PTEN-deficient B cells. SHIP is induced by ligation of the B cell inhibitory Fc receptor FcγRIIb. Co-clustering of the BCR and FcγRIIb, for example by immune complexes, strongly inhibits BCR signal strength.[12] Although SHIP has a predominantly inhibitory role $PI(3,4)P_2$ generated by SHIP contributes to activation of PKB.[28]

To date the study of PI3K signalling in lymphocytes has focussed mainly on the p110δ isoform as although other subunits are widely expressed throughout different tissue types the expression of p110δ appears to be concentrated in leucocytes.[6,34] However the exact tissue distribution of p110δ is not fully established, and it may be that p110δ has important functions outside the immune system. Indeed, expression of p110δ has recently been detected in neuronal tissue.[11] Conversely, it is important to note that p110δ is not the only catalytic isoform present in lymphocytes. Real-time PCR measurements in our laboratory have shown that p110α, p110δ and p110γ subunits are expressed at high levels in mature B cells although p110β is expressed at a much lower level (unpublished data). That said, currently most data exist about the role of p110δ in B cell signalling, and the induction of PIP_3 production is virtually absent in B cells lacking p110δ.[7] Similarly, there is a significant reduction in PKB phosphorylation in both resting and BCR-stimulated p110δ-deficient B cells.[7] This suggests that p110δ plays a crucial and non-redundant role in transducing BCR-mediated signals (Fig.1).

4   PI3K in B Cell Development

An understanding of the role of PI3K in B cell development is important as in many of the current mouse PI3K knockout models it can be difficult to disentangle the effects of PI3K in the response to antigen from its effects on earlier B cell development. Our current knowledge of PI3K in B cell development comes mainly from mice with germline deletion or mutation of components of the PI3K pathway. PI3K over-activity has also been studied in mice with B cell conditional deletion of PTEN (henceforth referred to as bPTEN).

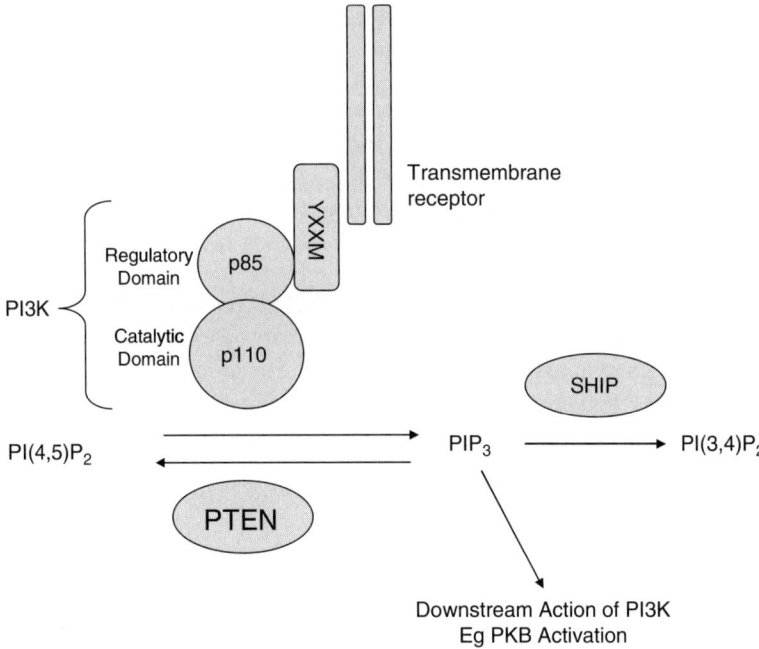

Fig. 1 Summary of PI3K action

Initial studies targeted the PI3K regulatory subunits. As p85α, p55α and p50α are splice variants of the same gene mouse strains have been created lacking all or some of these subunits. The most severe phenotype is seen in mice with deletion of all three splice variants with most dying within a few days of birth. When lymphocytes from these mice were examined in a Rag2$^{-/-}$ blastocyst complementation system total B cell number was greatly reduced with an apparent partial block of B cells at the pro-B to pre-B cell stage and a near complete absence of B1 cells.[13] This may reflect competition between p85-deficient pro-B cells and Rag2-deficient pro-B cells in the bone marrow. B1 cells are a specialised subset of mature B cells predominantly localised to the peritoneal and pleural cavities and spleen. Along with marginal zone (MZ) B cells they secrete much of the body's natural antibody and contribute to the primary humoral immune response to pathogens. Other mouse strains able to express some of the p85α splice variants, or mice with deletion of p85β alone, show a similar but less severe phenotype.[33] Those mice with a more severe phenotype also show a reduction in p110 catalytic subunit levels suggesting a stabilising effect of the regulatory subunit upon the catalytic subunit.[13]

P110α and p110β germline-targeted mice have also been created.[3,4] Both these have a lethal phenotype in utero. In contrast p110δ-targeted mice are viable. Two p110δ knockout lines and a kinase-dead knockin mutant have been created.[7,16,21]

All show a near complete absence of both B1 and marginal zone B cells with normal or mildly reduced numbers of follicular B cells. The fact that the phenotype of the p110δ knockout is less severe than that of the p85 regulatory subunit knockout suggests that catalytic subunits other than p110δ contribute to overall PI3K activity in B cell development. A viable p110γ knockout has also been reported but has no apparent B cell phenotype.[15]

The converse effects of PI3K over-activity upon B cell development have been reported in a B cell conditional PTEN-deleted (bPTEN) mouse model (germline deletion of PTEN is lethal early in utero[10]). In these mice an excess of B1 cells and MZ B cells is observed.[2,32] Mice with heterozygous deletion of PTEN show an intermediate number of B1 and MZ B cells (own unpublished observation).

In summary, current studies indicate that class Ia PI3K activity regulates development of B1 and MZ B cells. The effect seems 'dose dependent' with greater impairment of PI3K activity leading to a greater reduction in B1 and MZ B cell numbers. As MZ and B1 cells are functionally distinct from follicular B cells any gene targeting model attempting to assess the function of PI3K in mature B cells must take into account these differences in naïve peripheral B cells populations. Although targeting of different subunits of PI3K may generate different phenotypes it is unclear whether this reflects distinct roles for individual subunits or instead reflects the overall level of PI3K activity in the cell.

5   The B Cell Response to Antigen

Upon encounter with activating antigen B cells proliferate and may differentiate into extra-follicular antibody-secreting cells (ASCs) to provide an early source of low-affinity immunoglobulins. Alternatively, in the presence of T cell help, B cells may undergo further intense proliferation in specialist structures within secondary lymphoid tissue known as germinal centres.[8] Here cells also undergo hypermutation of the immunoglobulin variable region genes and selection for cells expressing high-affinity BCR. These processes are known as somatic hypermutation (SHM) and affinity maturation. Those cells with the highest-affinity BCR are selected to differentiate into high-affinity ASCs.[29] Prior to ASC differentiation B cells may also undergo switching of the immunoglobulin constant region genes to allow secretion of isotypes other than IgM – a process termed class switch recombination (CSR). To maintain immunity and prevent the development of autoimmunity it is critical that SHM and CSR are shut down once the B cell commits to ASC differentiation. Signalling via the BCR plays an important role in regulating these processes of proliferation, SHM, CSR and ASC differentiation.[27] This regulation likely depends upon PI3K activity downstream of the BCR.[25]

Most information regarding the activation and differentiation of mature B cells comes from analysis of the three p110δ-targeted mouse lines, from bPTEN conditional knockout mice and from the use of pharmacological inhibitors of PI3K. The interpretation of each suffers from potential difficulties: these include the complicating effects on B cell development, compensatory increases in other PI3K subunits, PI3K suppression in T cells, phosphatase-independent functions of PTEN, phosphatase activity of PTEN outside of the PI3K system and off target actions of pharmacological inhibitors.

6   Role of PI3K in the B Cell Response to Antigen

6.1   PI3K Promotes B Cell Proliferation Following BCR Ligation

In vitro p85α-deficient B cells show markedly reduced proliferation in response to stimuli including BCR ligation, CD40 ligation and LPS.[13,33] Similarly in p110δ-deficient B cells the proliferative response to BCR cross-linking is virtually abolished.[7,16,21] CD40 or LPS responses are impaired, but less severely. Pharmacological inhibitors of P110δ activity also strongly inhibit B cell proliferation.[5] This report also provided evidence that p110δ was activated by IL-4, a cytokine that plays a role in the responses of B cells to antigen and acts as a co-stimulus for in vitro proliferation. Conversely, both MZ and FO bPTEN cells showed increased proliferation in response to BCR cross-linking, anti-CD40 or LPS stimulation in vitro.[32] Despite this, it is unclear whether bPTEN cell hyperproliferate in vivo. Interestingly, despite the hyper-proliferative phenotype of bPTEN cells, B cell tumours are not observed in these mice. This is in contrast to PTEN deletion in other tissues such as T lymphocytes where tumour development is frequent. The reason for this is unclear.

Whilst the proliferative response to BCR, CD40 or LPS appears critically dependent on PI3K, proliferative responses to alternative stimuli, such as the combination of anti-CD40 and IL4, seem less dependent on PI3K implying the use alternative signalling pathways.[13]

6.2   The Humoral Immune Response is Suppressed
in the Absence of PI3K Activity

p110δ-deficient mice show reduced plasma levels of immunoglobulin.[7,16,22] The exact isotypic pattern of reduction varies between the mouse lines, but always results in lowered IgM. Conversely bPTEN mice show increased IgM.[32] As a majority of natural antibody is produced by B1 and MZ B cells it is not clear if this indicates a specific role for p110δ in immunoglobulin production or merely reflects the numbers of B1 and MZ B cells in these mice. All three p110δ-targeted mice also report reduced response of all immunoglobulin isotypes to both thymus-dependent and thymus-independent immunisation. In addition, germinal centres were reduced in size and number in p110δ-deficient mice although the possibility that this may relate to deficient PI3K activity in T cells or dendritic cells has not been excluded.

6.3   Class Switch Recombination is Inhibited by PI3K

Several lines of evidence suggest that the PI3K pathway inhibits CSR. bPTEN mice secrete normal or elevated levels of IgM in both the resting state and upon immunisation with thymus-dependent antigen. However, production of immunoglobulin of other isotypes is greatly reduced.[32] Similarly, when cultured in vitro PTEN-deficient B cells show greatly reduced ability to class switch and secrete minimal class switched immunoglobulin, despite producing normal levels of IgM.[24] Although one report describes absence of germinal centres in bPTEN mice immunised with NPCG other immunising agents appear to induce normal germinal centres yet still demonstrate suppression of CSR.[24] Thus suppression of CSR by PI3K in vivo does not appear to be a secondary consequence of lack of germinal centre formation. The suppression of CSR in bPTEN cells has been shown to be partially reversed in vitro by pharmacological inhibitors of p110δ such as IC87114.[24] This observation was somewhat surprising given an earlier study that suggested p110δ was downstream of IL-4, a cytokine that positively regulates CSR.[5] Perhaps PIP_3 levels need to be precisely controlled in time and space for CSR to function normally. The process of CSR relies critically upon the enzyme activation-induced cytidine deaminase (AID). In keeping with the reduced CSR real-time PCR measurements show reduced levels of AID mRNA in stimulated bPTEN B cells. Moreover, expression of post-switch circle transcripts which arise after CSR was reduced.[32] Expression in wild-type B cells of a constitutively active form of PKB reproduces the same suppression of both CSR and AID message level. Retroviral infection studies suggest that PKB-dependent suppression of CSR may be mediated partly by inactivation of the transcription factor FOXO1.[24] PI3K activity may also reduce CSR through its phosphorylation and cytoplasmic sequestration of the transcriptional repressor Bach2, as Bach2 knockout mice show reduced AID levels and reduced CSR.[18,19] Although the evidence for PI3K-mediated suppression of CSR is strong, these experiments on bPTEN mice do not exclude the possibility that PTEN also regulates CSR through a mechanism independent of its lipid phosphatase activity. PTEN has recently been shown to be essential for the maintenance of chromosomal integrity. This activity is independent of the action PTEN on PIP3 and instead relies on its ability to directly regulate the transcription of Rad51.[30] Rad51 mediates recombinational repair of double-stranded DNA damage; however, it is also required for the process of CSR.[26] It is not excluded that changes in Rad51 expression may contribute to the suppression of CSR seen in bPTEN mice.

Although bPTEN B cells show suppressed CSR B cells from p110δ-deficient mice do not show a significant enhancement of CSR above that seen in WT cells.[39] This is in contrast to reports of increased CSR in wild-type B cells cultured with p110δ inhibitor IC87114.[24] This discrepancy may reflect a greater suppression of overall PI3K activity in the presence of IC87114 than in the p110δ knockout due to compensatory effects of other catalytic isoforms in the knockout and suppression of other PI3K pathway components by the inhibitor.

While both SHM and CSR are mediated by AID the effect of PI3K on SHM has not been established.

6.4 Antibody Production and Plasma Cell Differentiation

Basal immunoglobulin levels are decreased in the p110δ- and p85-targeted mice and IgM levels are increased in bPTEN mice.[25] One published study of PI3K hyperactivity in bPTEN cells has shown increased ASC differentiation in vitro.[24] However these observations could reflect differences in the size of B1 and MZ B cell populations. These cells secrete natural IgM and are predisposed to ASC differentiation; therefore, comparisons of serum immunoglobulin levels or of ASC differentiation assays using unsorted B cells shed little light on the role of PI3K in the regulation of ASC differentiation and immunoglobulin production.[23] The Omori study also examined the effects of a p110δ selective inhibitor upon ASC differentiation, but at the dose used it is not possible to exclude non-specific toxic effects of the inhibitor.[24] Finally, in the same study, forced expression of activated PKB in B cells led to an insignificant enhancement of ASC differentiation. Despite the difficulty in interpreting the above observations, a role for PI3K in the promotion of ASC differentiation remains mechanistically plausible. The PI3K pathway is known to phosphorylate the transcriptional repressor Bach2 which in turn leads to its cytoplasmic sequestration.[38] A known target of Bach2-mediated repression is Blimp1, a transcription factor essential for ASC differentiation.[19] Thus, through its regulation of Bach2 activity PI3K could lead to elevated Blimp1 transcription and subsequently increased ASC differentiation.

In summary, although there is some suggestion that PI3K regulates ASC differentiation this remains to be proven. Further knockout models employing inducible Cre systems or studies performed on sorted B cells from bPTEN mice are awaited to resolve this issue.

7 Summary

The PI3K signalling pathway is crucial to normal B cell development and response to antigen. The p110δ catalytic subunit plays an important and non-redundant role within this pathway although other catalytic isoforms may also contribute. Although CD40, TLR and cytokines all activate PI3K the BCR seems especially dependent upon PI3K signalling. The downstream effects of PI3K may be mediated to a large extent by activation of PKB. In B cell development PI3K promotes development of MZ and B1 cells. In the response to antigen PI3K is crucial to BCR-mediated proliferation. PI3K activity has been shown to be inhibitory to CSR. The effects on immunoglobulin production and ASC differentiation are harder to disentangle from the developmental effects on cell populations and at present remain uncertain.

Acknowledgements We thank our colleagues for many stimulating discussions. DJH is a Cancer Research UK McElwain Clinical Research Training Fellow and MT is a Senior Non-Clinical Fellow of the Medical Research Council. Both are additionally supported by the Biotechnology and Biological Sciences Research Council.

References

1. Anderson, K.E., Coadwell, J., Stephens, L.R. & Hawkins, P.T. (1998) Translocation of PDK-1 to the plasma membrane is important in allowing PDK-1 to activate protein kinase B. *Curr Biol*, **8,** 684–691
2. Anzelon, A.N., Wu, H. & Rickert, R.C. (2003) Pten inactivation alters peripheral B lymphocyte fate and reconstitutes CD19 function. *Nat Immunol*, **4,** 287–294
3. Bi, L., Okabe, I., Bernard, D.J. & Nussbaum, R.L. (2002) Early embryonic lethality in mice deficient in the p110beta catalytic subunit of PI 3-kinase. *Mamm Genome*, **13,** 169–172
4. Bi, L., Okabe, I., Bernard, D.J., Wynshaw-Boris, A. & Nussbaum, R.L. (1999) Proliferative defect and embryonic lethality in mice homozygous for a deletion in the p110alpha subunit of phosphoinositide 3-kinase. *J Biol Chem*, **274,** 10963–10968
5. Bilancio, A., Okkenhaug, K., Camps, M., Emery, J.L., Ruckle, T., Rommel, C. & Vanhaesebroeck, B. (2006) Key role of the p110delta isoform of PI3K in B-cell antigen and IL-4 receptor signaling: comparative analysis of genetic and pharmacologic interference with p110delta function in B cells. *Blood*, **107,** 642–650
6. Chantry, D., Vojtek, A., Kashishian, A., Holtzman, D.A., Wood, C., Gray, P.W., Cooper, J.A. & Hoekstra, M.F. (1997) p110delta, a novel phosphatidylinositol 3-kinase catalytic subunit that associates with p85 and is expressed predominantly in leukocytes. *J Biol Chem*, **272,** 19236–19241
7. Clayton, E., Bardi, G., Bell, S.E., Chantry, D., Downes, C.P., Gray, A., Humphries, L.A., Rawlings, D., Reynolds, H., Vigorito, E. & Turner, M. (2002) A crucial role for the p110delta subunit of phosphatidylinositol 3-kinase in B cell development and activation. *J Exp Med*, **196,** 753–763
8. Cozine, C.L., Wolniak, K.L. & Waldschmidt, T.J. (2005) The primary germinal center response in mice. *Curr Opin Immunol*, **17,** 298–302
9. Del Nagro, C.J., Otero, D.C., Anzelon, A.N., Omori, S.A., Kolla, R.V. & Rickert, R.C. (2005) CD19 function in central and peripheral B-cell development. *Immunol Res*, **31,** 119–131
10. Di Cristofano, A., Pesce, B., Cordon-Cardo, C. & Pandolfi, P.P. (1998) Pten is essential for embryonic development and tumour suppression. *Nat Genet*, **19,** 348–355
11. Eickholt, B.J., Ahmed, A.I., Davies, M., Papakonstanti, E.A., Pearce, W., Starkey, M.L., Bilancio, A., Need, A.C., Smith, A.J., Hall, S.M., Hamers, F.P., Giese, K.P., Bradbury, E.J. & Vanhaesebroeck, B. (2007) Control of axonal growth and regeneration of sensory neurons by the p110delta PI 3-kinase. *PLoS ONE*, 2, e869
12. Fruman, D.A. (2004) Phosphoinositide 3-kinase and its targets in B-cell and T-cell signaling. *Curr Opin Immunol*, **16,** 314–320
13. Fruman, D.A., Snapper, S.B., Yballe, C.M., Davidson, L., Yu, J.Y., Alt, F.W. & Cantley, L.C. (1999) Impaired B cell development and proliferation in absence of phosphoinositide 3-kinase p85alpha. *Science*, **283,** 393–397
14. Hawkins, P.T., Anderson, K.E., Davidson, K. & Stephens, L.R. (2006) Signalling through class I PI3Ks in mammalian cells. *Biochem Soc Trans*, **34,** 647–662
15. Hirsch, E., Katanaev, V.L., Garlanda, C., Azzolino, O., Pirola, L., Silengo, L., Sozzani, S., Mantovani, A., Altruda, F. & Wymann, M.P. (2000) Central role for G protein-coupled phosphoinositide 3-kinase gamma in inflammation. *Science*, **287,** 1049–1053
16. Jou, S.T., Carpino, N., Takahashi, Y., Piekorz, R., Chao, J.R., Carpino, N., Wang, D. & Ihle, J.N. (2002) Essential, nonredundant role for the phosphoinositide 3-kinase p110delta in signaling by the B-cell receptor complex. *Mol Cell Biol*, **22,** 8580–8591
17. Maehama, T. & Dixon, J.E. (1998) The tumor suppressor, PTEN/MMAC1, dephosphorylates the lipid second messenger, phosphatidylinositol 3,4,5-trisphosphate. *J Biol Chem*, **273,** 13375–13378
18. Muto, A., Tashiro, S., Nakajima, O., Hoshino, H., Takahashi, S., Sakoda, E., Ikebe, D., Yamamoto, M. & Igarashi, K. (2004) The transcriptional programme of antibody class switching involves the repressor Bach2. *Nature*, **429,** 566–571

19. Ochiai, K., Katoh, Y., Ikura, T., Hoshikawa, Y., Noda, T., Karasuyama, H., Tashiro, S., Muto, A. & Igarashi, K. (2006) Plasmacytic transcription factor Blimp-1 is repressed by Bach2 in B cells. *J Biol Chem*, **281**, 38226–38234

20. Okada, T., Maeda, A., Iwamatsu, A., Gotoh, K. & Kurosaki, T. (2000) BCAP: the tyrosine kinase substrate that connects B cell receptor to phosphoinositide 3-kinase activation. *Immunity*, **13**, 817–827

21. Okkenhaug, K., Bilancio, A., Farjot, G., Priddle, H., Sancho, S., Peskett, E., Pearce, W., Meek, S.E., Salpekar, A., Waterfield, M.D., Smith, A.J. & Vanhaesebroeck, B. (2002) Impaired B and T cell antigen receptor signaling in p110delta PI 3-kinase mutant mice. *Science*, **297**, 1031–1034

22. Okkenhaug, K. & Vanhaesebroeck, B. (2003) PI3K in lymphocyte development, differentiation and activation. *Nat Rev Immunol*, **3**, 317–330

23. Oliver, A.M., Martin, F., Gartland, G.L., Carter, R.H. & Kearney, J.F. (1997) Marginal zone B cells exhibit unique activation, proliferative and immunoglobulin secretory responses. *Eur J Immunol*, **27**, 2366–2374

24. Omori, S.A., Cato, M.H., Anzelon-Mills, A., Puri, K.D., Shapiro-Shelef, M., Calame, K. & Rickert, R.C. (2006) Regulation of class-switch recombination and plasma cell differentiation by phosphatidylinositol 3-kinase signaling. *Immunity*, **25**, 545–557

25. Omori, S.A. & Rickert, R.C. (2007) Phosphatidylinositol 3-kinase (PI3K) signaling and regulation of the antibody response. *Cell Cycle*, **6**, 397–402

26. Petersen, S., Casellas, R., Reina-San-Martin, B., Chen, H.T., Difilippantonio, M.J., Wilson, P.C., Hanitsch, L., Celeste, A., Muramatsu, M., Pilch, D.R., Redon, C., Ried, T., Bonner, W.M., Honjo, T., Nussenzweig, M.C. & Nussenzweig, A. (2001) AID is required to initiate Nbs1/gamma-H2AX focus formation and mutations at sites of class switching. *Nature*, **414**, 660–665

27. Phan, T.G., Paus, D., Chan, T.D., Turner, M.L., Nutt, S.L., Basten, A. & Brink, R. (2006) High affinity germinal center B cells are actively selected into the plasma cell compartment. *J Exp Med*, **203**, 2419–2424

28. Scheid, M. (2002) PIP3 is essential but not sufficient for protein kinase B activation; PIP2 is required for PKB phosphorylation at ser-473. *J Biol Chem*, **277**, 9027–9035

29. Shapiro-Shelef, M. & Calame, K. (2005) Regulation of plasma-cell development. *Nat Rev Immunol*, **5**, 230–242

30. Shen, W.H., Balajee, A.S., Wang, J., Wu, H., Eng, C., Pandolfi, P.P. & Yin, Y. (2007) Essential role for nuclear PTEN in maintaining chromosomal integrity. *Cell*, **128**, 157–170

31. Suzuki, A., de la Pompa, J.L., Stambolic, V., Elia, A.J., Sasaki, T., del Barco Barrantes, I., Ho, A., Wakeham, A., Itie, A., Khoo, W., Fukumoto, M. & Mak, T.W. (1998) High cancer susceptibility and embryonic lethality associated with mutation of the PTEN tumor suppressor gene in mice. *Curr Biol*, **8**, 1169–1178

32. Suzuki, A., Kaisho, T., Ohishi, M., Tsukio-Yamaguchi, M., Tsubata, T., Koni, P.A., Sasaki, T., Mak, T.W. & Nakano, T. (2003) Critical roles of Pten in B cell homeostasis and immunoglobulin class switch recombination. *J Exp Med*, **197**, 657–667

33. Suzuki, H., Terauchi, Y., Fujiwara, M., Aizawa, S., Yazaki, Y., Kadowaki, T. & Koyasu, S. (1999) Xid-like immunodeficiency in mice with disruption of the p85alpha subunit of phosphoinositide 3-kinase. *Science*, **283**, 390–392

34. Vanhaesebroeck, B., Welham, M.J., Kotani, K., Stein, R., Warne, P.H., Zvelebil, M.J., Higashi, K., Volinia, S., Downward, J. & Waterfield, M.D. (1997) P110delta, a novel phosphoinositide 3-kinase in leukocytes. *Proc Natl Acad Sci USA*, **94**, 4330–4335

35. Vigorito, E., Bardi, G., Glassford, J., Lam, E.W., Clayton, E. & Turner, M. (2004) Vav-dependent and vav-independent phosphatidylinositol 3-kinase activation in murine B cells determined by the nature of the stimulus. *J Immunol*, **173**, 3209–3214

36. Vigorito, E. & Turner, M. (2006) Differential requirements of PI3K subunits for BCR or BCR/CD19-induced ERK activation. *Adv Exp Med Biol*, **584**, 43–52

37. Yamazaki, T., Takeda, K., Gotoh, K., Takeshima, H., Akira, S. & Kurosaki, T. (2002) Essential immunoregulatory role for BCAP in B cell development and function. *J Exp Med*, **195,** 535–545
38. Yoshida, C., Yoshida, F., Sears, D.E., Hart, S.M., Ikebe, D., Muto, A., Basu, S., Igarashi, K. & Melo, J.V. (2007) Bcr-Abl signaling through the PI-3/S6 kinase pathway inhibits nuclear translocation of the transcription factor Bach2, which represses the antiapoptotic factor heme oxygenase-1. *Blood*, **109,** 1211–1219
39. Janas, M., Hodson, D., et al. (2008) The effect of deleting p110delta on the phenotype and function of PTEN-deficient B cells. *J. Immunol.* **180(2)**,739–746

CD28: Old Dog, New Tricks

CD28 in Plasma Cell/Multiple Myeloma Biology

Jayakumar R. Nair, Cheryl Rozanski, and Kelvin P. Lee(✉)

1 Introduction

The concept of a costimulatory signal requirement for immune cell activation has been attributed to Bretscher and Cohn,[7] who first proposed that B cell activation required two signals. This model was subsequently modified by Lafferty and Cunningham for T cell activation and allograft rejection.[7,34] For naïve T cell activation, signal one is delivered by the T cell receptor (TCR) binding to cognate antigen presented by major histocompatibility complex (MHC) molecules on professional antigen presentation cells (APCs). The second signal is delivered by costimulatory receptor binding to its ligand(s) on the APC. While the costimulatory signal alone characteristically has no effect on T cells, in combination with a signal 1 it has been clearly shown to enhance cytokine secretion, proliferation, metabolic fitness, and survival during T cell activation.

The prototypic costimulatory receptor is CD28, and its effect on T cell activation has been extensively characterized. However, CD28 was also initially described to be expressed on normal plasma cells but not B cells,[29] and subsequently on the transformed counterparts of these cells (namely, multiple myeloma) – where it has been shown to be significantly associated with disease progression and poor prognosis.[14,52,56] However, plasma cells (both normal and transformed) no longer express an antigen-specific receptor to deliver a signal 1, and the biological role of CD28 in the B cell lineage has been unclear. Yet CD28 expression is highly regulated (suppressed) in B cells by Pax5, a key transcriptional regulator of mature B cell → plasma cell differentiation,[13] supporting a role for CD28 in plasma cell biology.

Because multiple myeloma (MM) cells (like normal PC) are critically dependent on (largely undefined) cell–cell interactions with the bone marrow microenvironment, these interactions are central in developing new therapeutic targets for this incurable disease. This review will examine the role of CD28 expressed on PC/MM cells as a key component of this interaction.

Kelvin P. Lee
Department of Immunology, Roswell Park Cancer Institute, Buffalo, NY 14263, USA

S.P. Schoenberger et al. (eds.) *Crossroads between Innate and Adaptive Immunity II*,
doi: 10.1007/978-0-387-79311-5_6, © Springer Science + Business Media, LLC 2009

2   Old Dog: CD28 and T Cells

CD28 has been most extensively characterized in the context of T cell biology. It was first described as a homodimeric T cell surface antigen consisting of 44-kDa subunits (initially called Tp44),[42] expressed on virtually all human CD4[+] and CD8[+] T cells,[27] as well as activated neutrophils,[66] bone marrow stromal cells involved in B lymphopoiesis,[20] and normal/transformed plasma cells (i.e., multiple myeloma cells).[29,52] Previous work in our lab demonstrated that CD28 mRNA expression in the myeloma cell line RPMI 8226 was identical to that of T cells (namely four splice variants of 1.3, 1.5, 3.5, and 3.7 kb), indicating the MM cells are not expressing a unique form of CD28.[37]

Cloning of the CD28 gene revealed it to be a member of the immunoglobulin gene (Ig) superfamily.[3] The CD28 receptor family itself contains a second member, CTLA4,[8] which shares significant amino acid identity with CD28, maps to the same location on chromosome 2 and binds to the same ligands (although CTLA4 binds with higher affinity[27,39]). CTLA4 however functions antagonistically to CD28, inhibiting T cell activation.[67] The ligand for CD28 and CTLA4 are CD80 (B7–1) and CD86 (B7–2), which are expressed predominantly on professional antigen-presenting cells (APCs), and in particular dendritic cells (DCs).[63]

Functionally, in T cells it was first shown that monoclonal antibody activation of CD28 plus TCR/CD3 crosslinking induces a 5–50-fold enhancement in the expression and secretion of IL-2, TNFα, lymphotoxin, IFN-γ, and GM-CSF,[64] in part via stabilization of these genes' mRNA transcripts.[38] Costimulation of CD28 was distinct from TCR/CD3-alone induced T cell activation, as the CD28 activation conferred resistance to the immunosuppressive effects of cyclosporine A.[26] Subsequent studies have shown that CD28 costimulation in T cells results in augmented T cell proliferation, effector function,[38,61] enhanced survival via upregulation of antiapoptotic gene bcl-x$_L$ (Boise 1995), more efficient glucose metabolism,[19] and the prevention of anergy.[58]

3   New Tricks: CD28 and Plasma Cells

While CD28 is classically thought of as a receptor that modulates T cell biology, it is also expressed on normal and transformed plasma cells (both murine and human). What role CD28 plays in plasma cell biology is almost entirely undefined. We initially found that CD28-deficient mice had lower constitutive levels of all antibody isotypes (including IgM), and had poor antibody responses following vaccination.[24] This was initially thought to be due to the lack of CD4 T cell help (although that does not necessarily explain the low IgM levels). However, it has been recently demonstrated in B cells that Pax5, a key regulator that inhibits B cell differentiation to plasma cells, suppresses CD28 expression as one of its target genes.[13] In addition, transplantation of CD28-deficient bone marrow into mice with normal T cell function also resulted in significantly lower antibody titers following vaccination, suggesting a primary role for CD28 in normal plasma cell biology.

3.1   Expression of CD28 and CD86 in Multiple Myeloma

What specific role CD28 plays in normal plasma cell biology remains unclear. However, observations in patients with multiple myeloma offer clues. Multiple myeloma (MM) is a treatable but incurable progressive hematologic malignancy characterized by the accumulation of neoplastic monoclonal plasma cells in the bone marrow.[2,22] Myeloma cells are characterized by a low proliferative index, enhanced cell survival, and resistance to chemotherapeutic drugs with a very high relapse rates.[33,41] It is thought that myeloma cell interactions with the bone marrow microenvironment, specifically bone marrow stromal cells (BMSCs), are essential for activation of prosurvival signaling pathways in the MM cells, as well as induction of cytokine and growth factor expression on the BMSC side[6,12,62,65] (Fig. 1). It is clear that for the majority of the disease course, multiple myeloma cells have same reliance on the bone marrow (BM) microenvironment as do normal PC. Thus, it might be expected that treatment pressure may select for expression of MM receptors involved in prosurvival interactions with the BM microenvironment. This is the pattern seen for CD28.

In MM patients, CD28 expression correlates significantly with disease progression – such that 26% of patients are highly positive at diagnosis, 59% at medullary relapse, 93% at extramedullary relapse, and 100% of secondary plasma cell leukemias (as well as almost all MM cell lines).[56] Other studies have demonstrated that CD28 expression prognostically correlates with a worse outcome following high dose chemotherapy plus autologous stem cell rescue.[1,43] All these findings point to a prosurvival role for CD28 in MM, where treatment pressure selects subsets of myeloma cells with high CD28 expression.

Somewhat unexpectedly, CD86 expression has also been found on 54% of MM patients at time of diagnosis, and also is associated with significantly poorer prognosis.[53] Whether this has anything to do with concurrent CD28 expression was not examined. However, taking the all the clinical studies in aggregate, we predict that all CD28+ myelomas are also CD86+, and we and others have found this to be true for myeloma cell lines.[4,28,69] This raises the possibility that myeloma cells themselves may engage in a CD28–CD86 "autocrine" prosurvival interaction, and this interaction may contribute to the progression to bone marrow stromal-independent myeloma – a significantly worse disease. Whether CD86 is also expressed on normal plasma cells has not been well studied.

3.2   CD28 Signaling Pathways in Myeloma Cells: Barking Up the Same Tree

In T cells receiving a signal 1, CD28 binding to CD80/CD86 leads predominantly to the activation of PI3K and MEKK pathways, both of which regulate NF-κB activity.[57,63] In myeloma cells, Zhang et al.[69] demonstrated that the p85 subunit of PI3K also associated with the cytoplasmic tail of CD28 when activated by CD80 binding.

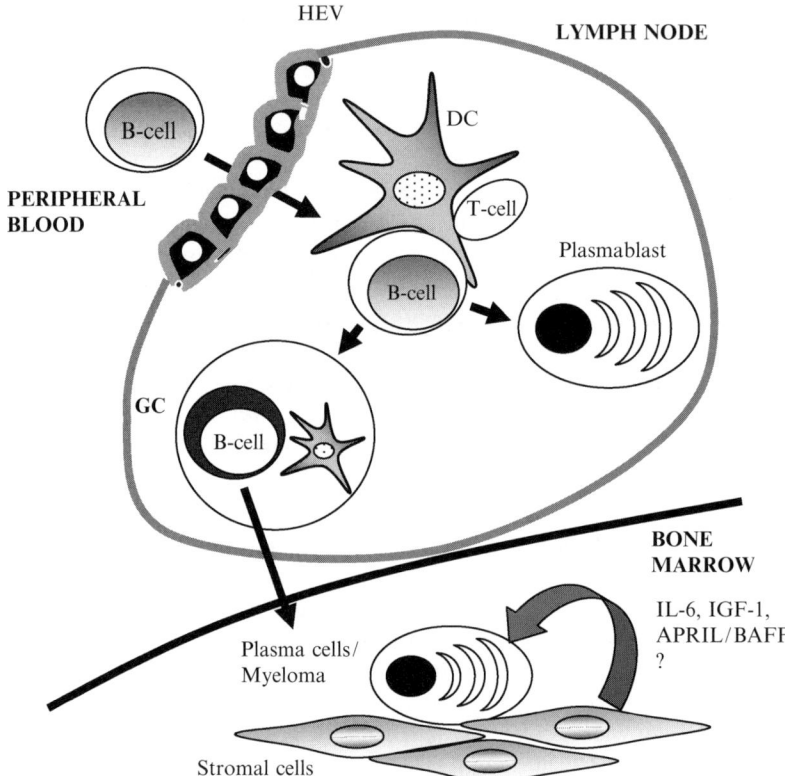

Fig. 1 A representation of the activation and terminal differentiation of B cells to form plasma cells and bone marrow-associated myeloma cells. Mature B cells traverse the high endothelial venules (HEV) and migrate into lymph nodes as do T cells. There the interdigitating dendritic cells (DCs) act as a matrix in which the Ag-specific T cells activate antigen-specific naïve B cells which differentiate into either short lived plasmablasts or germinal center B cells. In GC, the B cells interact with follicular DCs and begin monoclonal expansion; undergo multiple rounds of VDJ hypermutations and class switching and antigen selection. Those that express high-affinity Ag receptors differentiate into long-lived plasmablasts/plasma cells that typically migrate to the bone marrow where they remain as nonproliferative and dormant. Myeloma cells are the transformed counterparts of the post-GC terminally differentiated plasma cells and localize to the bone marrow where they interact with stromal cells via adhesion molecules and induce the secretion of growth factors such as IL-6 to aid in its survival and proliferation

Consistent with this observation, our studies have shown that direct activation of CD28 by anti-CD28 monoclonal Ab *alone* (without a signal 1) resulted in rapid PI3K signaling in the three human myeloma cell lines (8226, U266, and MM1S) studied, with sustained levels up to 24 h.[4] Downstream of PI3K activation we found reduction of total IκBα, increases in total NF-κB signaling (as measured by electromobility gel shift assays and upregulation of NF-κB target genes), and increased signaling through canonical (p65, p50/p52 heterodimers) and noncanonical (RelB) pathways. Importantly, NF-κB signaling in myeloma cells has been clearly shown to be prosurvival.

Why CD28 activation by itself in myeloma cell results in detectable signaling, in contrast to T cells where a signal 1 is also clearly required, is currently unclear. Other groups have not been able to identify a signal 1 that is costimulated by CD28 activation in myeloma cells, except possibly protein kinase C activation for IL-8 production.[62] While it is formally possible that some unknown signal 1 is being delivered, we have seen robust CD28-induced NF-κB signaling in the complete absence of serum – indicating that exogenously added soluble factors are unlikely to be playing a role. One possibility for the nondependence on a signal 1 is that inhibition of PI3K signaling (e.g., via PTEN) is more dominant in T cells than in myeloma cells. This lower threshold may allow for transmission of a CD28 signal in myeloma cells in the absence of other synchronous signals.

3.3   CD28 and Multiple Myeloma: Cellular Responses

A hallmark of CD28 costimulation in T cells is the markedly augmented cytokine production, in particular IL-2. Although CD28 expression on MM cells is similar to that observed in T cells, CD28 activation does not upregulate IL-2 production in myeloma cells. However, Shapiro et al.[62] have shown that the promoter of the proangiogenic cytokine IL-8 contains a CD28 response element (CD28 RE) and that MM cell production of IL-8 is upregulated by CD28 activation.

In T cells, activation of CD28 in conjunction with a signal one (but not by itself) augments cell proliferation. Somewhat unexpectedly, in myeloma cells we found that CD28 activation (by anti-CD28 mAb or CD80/CD86 binding) by itself inhibited cell proliferation by ~50% (via a G_0/G_1 cell cycle arrest) without inducing apoptosis.[4] The reason for this differential response compared to T cells is currently unclear, but certainly the autocrine production of proproliferative cytokines in T cells (namely IL-2) is not seen in MM cells. For example, IL-6 production is not induced by CD28 activation in MM cells.[62] It is also possible that CD28 signaling in myeloma cells is not as "complete" as it is in T cells. In T cells, CD28 signaling can be divided into two independent downstream pathways – with PI3K being responsible for survival and viability, while the factors that bind to the C-terminal proline motifs affect proliferation and cytokine responses.[9] We cannot rule out the possibility that this latter proliferative signaling pathway is inactive in MM cells.

Similar to T cells, however, we have found that CD28 activation in myeloma cells (both cell lines and primary patient isolates) increases survival (~2-fold) when cells are exposed to apoptosis-inducing conditions, including serum starvation and dexamethasone (which is widely used as a chemotherapeutic agent in this disease).[4] This is not however through the upregulation of Bcl-2 or Bcl-x_L. Other groups using a different anti-CD28 mAb also found that CD28 activation inhibited MM cell proliferation, but also induced apoptosis.[54] This difference may be explained by sensitivity of different MM cell lines to the disruption of the myeloma–myeloma CD28–CD86 interaction (see later).

Interestingly, many of the cellular responses to CD28 activation are the same as those elicited by integrin-mediated adhesion of myeloma cells to extracellular matrix (ECM), which has been called cell adhesion-mediated drug resistance (CAM-DR) by Dalton and colleagues.[11,23,35] These responses include induction of NF-κB signaling, decreased proliferation and a G_1 arrest cell cycle arrest, and enhanced survival/resistance to chemotherapeutic agents. It is also possible that there is crosstalk between integrin and CD28 signaling, as it has been shown that the CD28-negative MM cell line MOLP-8 can be induced to express CD28 by culture on bone marrow stromal feeder cells (and presumably the ECM made by these cells) in vitro.[44]

4   CD28 and the Microenvironment

If CD28 is a prosurvival receptor for myeloma cells, how is it being activated in the myeloma microenvironment? Ligand-mediated CD28 activation is thought to occur primarily in the context of direct cell–cell contact with CD80/CD86+ cells. In the case of T cells, these cells are APCs, and in particular DCs. In the case of myeloma cells, a second possibility exists, namely, CD86 expressed on other myeloma cells.

4.1   Myeloma–Myeloma Interaction

While a direct assessment of CD28 and CD86 coexpression on primary myeloma cells has not been done in a comprehensive fashion, clinical studies examining the individual expression of CD28 or CD86 lead us to conclude that most (if not all) CD28+ myelomas coexpress CD86. We have found this to be the case in the three CD28+ MM cell lines we examined (RPMI 8226, U266, and MM1.S), which is consistent with other reports examining other MM cell lines.[28,69] None of the cell lines expressed CD80 or CTLA4 (by FACS).

It has not been previously reported whether a CD28–CD86 myeloma–myeloma interaction affects MM cell biology in vivo. The possibility of autocrine CD28–CD86 activation is supported by some,[28] but not all,[69] in vitro studies. Our initial findings indicate that blocking this interaction with the soluble fusion receptor CTLA4-Ig (abatacept, which has been FDA approved for use in rheumatoid arthritis) that binds to CD80/CD86 results in downregulation of constitutive NF-κB signaling, direct induction of apoptosis in some cell lines, and sensitization of all cell lines to chemotherapy. These findings point to a prosurvival role for this interaction, and raise the intriguing possibility that this MM–MM interaction plays a role in the progression to stromal independence, replacing some supportive interactions with the bone marrow microenvironment with an "autocrine" CD28–CD86 interaction. If true, this interaction would not be easily dispensed with by the myeloma cells and would represent an attractive therapeutic target (for example, combining CTLA4-Ig with chemotherapy).

4.2 Myeloma–DC Interactions

We have found that that even with ongoing myeloma CD28 activation by myeloma CD86, this signal could be greatly increased by anti-CD28 mAb treatment – indicating that the autocrine interaction only generates a minority of the total potential CD28 signal.[4] This is likely due to the relative inefficiency of myeloma cells adhering to one another to form stable interactions. One CD80/CD86+ cell that is particularly adept at interacting with other immune cells is the dendritic cell, which for T cells is the major "source" of CD28 ligands in vivo.

There is considerable evidence that DCs are directly involved in the survival, proliferation, and differentiation of normal B cells and plasma cells. Dendritic cells produce IL-6,[60,70] and we have found that myeloma cells can induce DC production of IL-6 via a CD28–CD80/CD86 interaction (see below). Dendritic cells and B cells form clusters in vitro and in vivo;[32] this direct interaction provides the B cells with proliferation and survival signals.[68] These DC-B cell interactions in the germinal center induce dramatic B cell expansion (via CD40 and IL-12) and drive plasma cell differentiation.[15,16,18] Additional studies have demonstrated that interactions with DC enhance plasmablast survival and differentiation in a T-independent response, in part through secretion of APRIL and/or BAFF.[5]

In multiple myeloma, a role for dendritic cells as part of the *prosurvival* tumor microenvironment is only beginning to be appreciated. However, it has been noted in murine models using implanted plasmacytomas that there is rapid tumor infiltration by substantial numbers of host conventional dendritic cells (and other APC).[10] Consistent with these findings, we and others have reported that in patient bone marrow biopsy specimens there is extensive and specific infiltration of conventional DC into myeloma infiltrates (18–79-fold more DC in myeloma infiltrates compared to the adjacent normal bone marrow).[4,55,59] It has been also shown that APRIL and BAFF protect myeloma cells against apoptosis caused by IL-6 withdrawal and dexamethasone,[48] and recently it has been shown that DC support the clonogenicity of primary human MM cells in vitro, in part via APRIL- and RANK-mediated interactions.[30]

We have also reported that DC and MM cells cluster together when cocultured in vitro.[4] Direct cell–cell contact-induced downregulation of myeloma cell proliferation (~50%), and this effect could be partially abrogated by either blocking the CD28–CD80/CD86 interaction (using CD28-Ig to bind to the B7 molecules) or coculturing the MM cells with immature DC expressing low levels of CD80/CD86. Importantly, coculture with dendritic cells protected the myeloma cells from dexamethasone-induced death.

4.3   DC–Myeloma Interactions

Not only do the supportive stromal cells of the MM microenvironment modulate the myeloma cells they interact with, but they themselves are modulated by this interaction. It has been shown for normal plasma cells and myeloma cells that coculture with a mixed population of BMSCs induces stromal cell production of IL-6.[40,47] Which stromal cell(s) is making the IL-6 and how it is induced to do so remains unanswered, but other soluble factors (such as TNFα, TGFβ, and VEGF) have been implicated in the induction of IL-6 in BMSC. In this regard, a significant aspect of dendritic cells as potential components of the myeloma microenvironment is their ability to secrete cytokines that support the survival of myeloma cells, including IL-6 and APRIL/BAFF.

Interestingly, it has been shown that IL-6 can be induced in dendritic cells in vitro by the soluble fusion receptor CD28-Ig via "back signaling" through CD80/CD86.[51] This raises the possibility that engagement of CD28 on myeloma cells to CD80/CD86 on DC not only triggers a CD28 prosurvival signal to the myeloma cell, but also a CD80/CD86-mediated signal to the DC, inducing the production of IL-6 (Fig. 2). Consistent with this, in initial in vitro studies we have found that coculturing myeloma cells with conventional dendritic cells induces a >1,000-fold increase in DC production of secreted IL-6 (from <7.5 to 23,000 pg ml^{-1}). This finding suggests that DC in myeloma microenvironment could be an abundant source of IL-6 and that induction of this cytokine is mediated (at least in part) via a myeloma CD28–DC CD80/86 interaction.

A second dendritic cell response to back signaling through CD80/CD86 (via CTLA4-Ig crosslinking instead of CD28-Ig, or by T cells) is induction of the tryptophan-catabolizing enzyme indoleamine 2,3 dioxygenase (IDO).[25,36,46,49,50] This enzyme catabolizes the essential amino acid tryptophan to kynurenine (and can be inhibited by 1-methyl-tryptophan (1-MT)), and depletion of tryptophan from the microenvironment significantly inhibits T cell activation.[17,21,25,45,49] Given that multiple myeloma patients have significant defects in cell-mediated immunity (which is a significant obstacle to effective immunotherapeutic strategies in this disease), it is possible that a myeloma CD28–DC CD80/86 interaction also induces DC production of IDO (Fig. 2). And in initial in vitro studies we have found that coculture of myeloma cells with conventional DC induces significant upregulation in the DC of IDO protein expression as well as enzymatic activity in a CD28-dependent manner.

Together, our early findings suggest that dendritic cells may not be simply a passive source of ligands to activate CD28 on myeloma cells. Rather, the DC is an active partner in this cellular interaction, and is induced to produce prosurvival and immunosuppressive factors. It is not unreasonable to believe this interaction is a "normal" one between normal plasma cells and bone marrow dendritic cells. DC production of IL-6 would support the survival of long-lived PC, and production of IDO would generate a bubble of local immunosuppression – preventing T cells that can recognize the new immunoglobulin idiotype being presented by the DC (in its

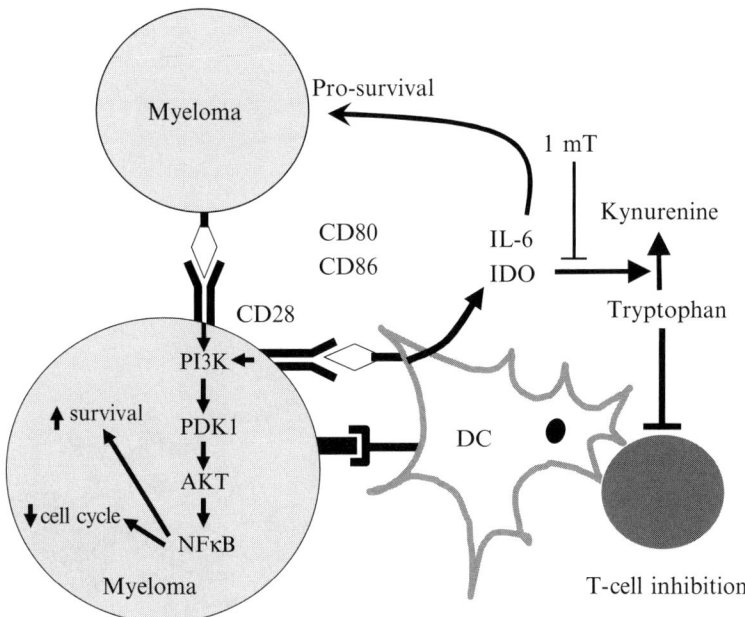

Fig. 2 Model of prosurvival signaling via CD28–CD86 interactions. Myeloma cells interact with themselves in an autocrine manner via CD28–CD86 interactions to induce NF-κB via the PI3K > PDK1 > AKT > NF-κB pathway. Upregulation of NF-κB is accompanied by increased survival against apoptotic agents, but lower proliferation. A second CD28–CD86 interaction occurs between MM cells and infiltrating DC. In addition to "forward" signaling of CD28 to the myeloma cell, "back signaling" through CD80/CD86 on the DC induces DC production of the prosurvival cytokine IL-6 and the tryptophan-catabolizing enzyme indoleamine 2,3 dioxygenase (IDO). IDO, by depleting microenvironmental tryptophan, inhibits T cell activation

APC function) from being activated. In myeloma, this is simply amplified by the much greater number of MM cells.

5   Therapeutically Targeting the CD28 Interaction in Multiple Myeloma

If the interaction of CD28 on the myeloma cell with the microenvironment plays an important role in the survival of these cells, especially under treatment pressure, then this interaction is an attractive therapeutic target. A variety of approaches can be envisioned (Fig. 3). CD28 itself (which cannot be readily downregulated by the MM cells) could be directly targeted with toxin/radionuclide-conjugated anti-CD28 mAb. It is likely that normal T cells would also be targeted by this approach, but

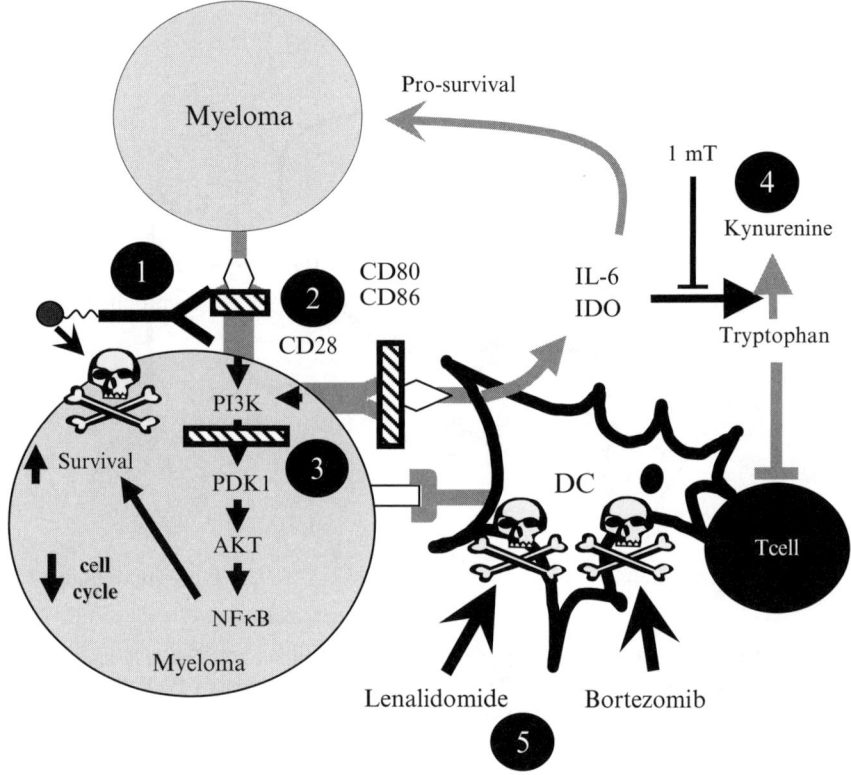

Fig. 3 Targeting the CD28 interaction in multiple myeloma. Potential therapeutic approaches for targeting the CD28 interaction in MM include: *1* direct targeting with toxin/radionuclide-conjugated anti-CD28 mAb, *2* blocking the CD28–CD80/CD86 interaction with fusion receptors (e.g., CTLA4-Ig) or antibody Fab fragments, *3* inhibiting signaling pathways downstream of the CD28 receptor, *4* blocking factors secreted by the DC, such as using 1-MT to inhibit IDO activity, and *5* targeting the DC

they could potentially be harvested and stored for reinfusion after the antibody treatment. A second approach is to block the CD28–CD80/CD86 interaction. We have initial preliminary data that CTLA4-Ig blockade can sensitize MM cells to chemotherapy, and this agent is already FDA approved. The ability of CTLA4-Ig itself to induce IDO production in DC is a confounding factor, but may be overcome by simultaneous treatment with IDO inhibitors such as 1-MT. A third approach would be to inhibit the signaling pathways that are downstream of the CD28 receptor. This is an area that is being actively pursued – not for multiple myeloma treatment, but for treatment of solid organ transplant rejection. In this regard, it is interesting to note that many agents used for transplant immunosuppression (e.g., rapamycin) also have activity against multiple myeloma. A fourth approach is to target the companion dendritic cell. Our initial studies suggest that lenalidomide, which has significant clinical activity via targeting the multiple myeloma microenvironment, significantly disrupts DC differentiation in vitro. Similarly, Kukreja et al.[31] have shown that bortezomib, which also is very active against multiple myeloma, disrupts myeloma cell-DC prosurvival interactions by killing the DC. Altogether, identification of CD28 as an important prosurvival receptor for myeloma cells opens a gateway to new treatment approaches for this disease.

References

1. Almeida, J., Orfao, A., Ocqueteau, M., Mateo, G., Corral, M., Caballero, M. D., Blade, J., Moro, M. J., Hernandez, J. and San Miguel, J. F., 1999, High-sensitive immunophenotyping and DNA ploidy studies for the investigation of minimal residual disease in multiple myeloma, *Br J Haematol* **107**(1):121–131
2. Anderson, K. C., Kyle, R. A., Dalton, W. S., Landowski, T., Shain, K., Jove, R., Hazlehurst, L. and Berenson, J., 2000, Multiple myeloma: new insights and therapeutic approaches, *Hematology (Am Soc Hematol Educ Program)* 147–165
3. Aruffo, A. and Seed, B., 1987, Molecular cloning of a CD28 cDNA by a high-efficiency COS cell expression system, *Proc Natl Acad Sci USA* **84**(23):8573–8577
4. Bahlis, N. J., King, A. M., Kolonias, D., Carlson, L. M., Liu, H. Y., Hussein, M. A., Terebelo, H. R., Byrne, G. E., Jr., Levine, B. L., Boise, L. H. and Lee, K. P., 2007, CD28-mediated regulation of multiple myeloma cell proliferation and survival, *Blood* **109**(11):5002–5010
5. Balazs, M., Martin, F., Zhou, T. and Kearney, J., 2002, Blood dendritic cells interact with splenic marginal zone B cells to initiate T-independent immune responses, *Immunity* **17**(3):341–352
6. Bisping, G., Leo, R., Wenning, D., Dankbar, B., Padro, T., Kropff, M., Scheffold, C., Kroger, M., Mesters, R. M., Berdel, W. E. and Kienast, J., 2003, Paracrine interactions of basic fibroblast growth factor and interleukin-6 in multiple myeloma, *Blood* **101**(7):2775–2783
7. Bretscher, P. and Cohn, M., 1970, A theory of self-nonself discrimination, *Science* **169**(950):1042–1049
8. Brunet, J. F., Denizot, F., Luciani, M. F., Roux-Dosseto, M., Suzan, M., Mattei, M. G. and Golstein, P., 1987, A new member of the immunoglobulin superfamily–CTLA-4, *Nature* **328**(6127):267–270
9. Burr, J. S., Savage, N. D., Messah, G. E., Kimzey, S. L., Shaw, A. S., Arch, R. H. and Green, J. M., 2001, Cutting edge: distinct motifs within CD28 regulate T cell proliferation and induction of Bcl-XL, *J Immunol* **166**(9):5331–5335
10. Corthay, A., Skovseth, D. K., Lundin, K. U., Rosjo, E., Omholt, H., Hofgaard, P. O., Haraldsen, G. and Bogen, B., 2005, Primary antitumor immune response mediated by CD4+ T cells, *Immunity* **22**(3):371–383
11. Damiano, J. S., Cress, A. E., Hazlehurst, L. A., Shtil, A. A. and Dalton, W. S., 1999, Cell adhesion mediated drug resistance (CAM-DR): role of integrins and resistance to apoptosis in human myeloma cell lines, *Blood* **93**(5):1658–1667
12. Dankbar, B., Padro, T., Leo, R., Feldmann, B., Kropff, M., Mesters, R. M., Serve, H., Berdel, W. E. and Kienast, J., 2000, Vascular endothelial growth factor and interleukin-6 in paracrine tumor-stromal cell interactions in multiple myeloma, *Blood* **95**(8):2630–2636
13. Delogu, A., Schebesta, A., Sun, Q., Aschenbrenner, K., Perlot, T. and Busslinger, M., 2006, Gene repression by Pax5 in B cells is essential for blood cell homeostasis and is reversed in plasma cells, *Immunity* **24**(3):269–281
14. Drexler, H. G. and Matsuo, Y., 2000, Malignant hematopoietic cell lines: in vitro models for the study of multiple myeloma and plasma cell leukemia, *Leukemia Res* **24**(8):681–703

15. Dubois, B., Barthelemy, C., Durand, I., Liu, Y. J., Caux, C. and Briere, F., 1999, Toward a role of dendritic cells in the germinal center reaction: triggering of B cell proliferation and isotype switching, *J Immunol* **162**(6):3428–3436

16. Dubois, B., Massacrier, C., Vanbervliet, B., Fayette, J., Briere, F., Bancherecau, J. and Caux, C., 1998, Critical role of IL-12 in dendritic cell-induced differentiation of naive B lymphocytes, *J Immunol* **161**(5):2223–2231

17. Fallarino, F., Vacca, C., Orabona, C., Belladonna, M. L., Bianchi, R., Marshall, B., Keskin, D. B., Mellor, A. L., Fioretti, M. C., Grohmann, U. and Puccetti, P., 2002, Functional expression of indoleamine 2,3-dioxygenase by murine CD8 alpha(+) dendritic cells, *Int Immunol* **14**(1):65–68

18. Fayette, J., Durand, I., Bridon, J. M., Arpin, C., Dubois, B., Caux, C., Liu, Y. J., Bancherecau, J. and Briere, F., 1998, Dendritic cells enhance the differentiation of naive B cells into plasma cells in vitro, *Scand J Immunol* **48**(6):563–570

19. Frauwirth, K. A., Riley, J. L., Harris, M. H., Parry, R. V., Rathmell, J. C., Plas, D. R., Elstrom, R. L., June, C. H. and Thompson, C. B., 2002, The CD28 signaling pathway regulates glucose metabolism, *Immunity* **16**(6):769–777

20. Gray Parkin, K., Stephan, R. P., Apilado, R. G., Lill-Elghanian, D. A., Lee, K. P., Saha, B. and Witte, P. L., 2002, Expression of CD28 by bone marrow stromal cells and its involvement in B lymphopoiesis, *J Immunol* **169**(5):2292–2302

21. Grohmann, U., Orabona, C., Fallarino, F., Vacca, C., Calcinaro, F., Falorni, A., Candeloro, P., Belladonna, M. L., Bianchi, R., Fioretti, M. C. and Puccetti, P., 2002, CTLA-4-Ig regulates tryptophan catabolism in vivo, *Nat Immunol* **3**(11):1097–1101

22. Hallek, M., Bergsagel, P. L. and Anderson, K. C., 1998, Multiple myeloma: increasing evidence for a multistep transformation process, *Blood* **91**(1):3–21

23. Hazlehurst, L. A., Damiano, J. S., Buyuksal, I., Pledger, W. J. and Dalton, W. S., 2000, Adhesion to fibronectin via beta1 integrins regulates p27kip1 levels and contributes to cell adhesion mediated drug resistance (CAM-DR), *Oncogene* **19**(38):4319–4327

24. Horspool, J. H., Perrin, P. J., Woodcock, J. B., Cox, J. H., King, C. L., June, C. H., Harlan, D. M., St Louis, D. C. and Lee, K. P., 1998, Nucleic acid vaccine-induced immune responses require CD28 costimulation and are regulated by CTLA4, *J Immunol* **160**(6):2706–2714

25. Hwu, P., Du, M. X., Lapointe, R., Do, M., Taylor, M. W. and Young, H. A., 2000, Indoleamine 2,3-dioxygenase production by human dendritic cells results in the inhibition of T cell proliferation, *J Immunol* **164**(7):3596–3599

26. June, C. H., Ledbetter, J. A., Gillespie, M. M., Lindsten, T. and Thompson, C. B., 1987, T-cell proliferation involving the CD28 pathway is associated with cyclosporine-resistant interleukin 2 gene expression, *Mol Cell Biol* **7**(12):4472–4481

27. June, C. H., Vandenberghe, P. and Thompson, C. B., 1994, The CD28 and CTLA-4 receptor family, *Chem Immunol* **59**:62–90

28. Kornbluth, J., 1995, Potential role of CD28-B7 interactions in the growth of myeloma plasma cells, *Curr Top Microbiol Immunol* **194**:43–49

29. Kozbor, D., Moretta, A., Messner, H. A., Moretta, L. and Croce, C. M., 1987, Tp44 molecules involved in antigen-independent T cell activation are expressed on human plasma cells, *J Immunol* **138**(12):4128–4132

30. Kukreja, A., Hutchinson, A., Dhodapkar, K., Mazumder, A., Vesole, D., Angitapalli, R., Jagannath, S. and Dhodapkar, M. V., 2006, Enhancement of clonogenicity of human multiple myeloma by dendritic cells, *J Exp Med* **203**(8):1859–1865

31. Kukreja, A., Hutchinson, A., Mazumder, A., Vesole, D., Angitapalli, R., Jagannath, S., O'Connor O, A. and Dhodapkar, M. V., 2007, Bortezomib disrupts tumour-dendritic cell interactions in myeloma and lymphoma: therapeutic implications, *Br J Haematol* **136**(1):106–110

32. Kushnir, N., Liu, L. and MacPherson, G. G., 1998, Dendritic cells and resting B cells form clusters in vitro and in vivo: T cell independence, partial LFA-1 dependence, and regulation by cross-linking surface molecules, *J Immunol* **160**(4):1774–1781

33. Kyle, R. A. and Rajkumar, S. V., 2004, Multiple myeloma, *N Engl J Med* **351**(18): 1860–1873

34. Lafferty, K. J., Warren, H. S., Woolnough, J. A. and Talmage, D. W., 1978, Immunological induction of T lymphocytes: role of antigen and the lymphocyte costimulator, *Blood Cells* **4**(3):395–406

35. Landowski, T. H., Olashaw, N. E., Agrawal, D. and Dalton, W. S., 2003, Cell adhesion-mediated drug resistance (CAM-DR) is associated with activation of NF-kappa B (RelB/p50) in myeloma cells, *Oncogene* **22**(16):2417–2421

36. Lee, J. R., Dalton, R. R., Messina, J. L., Sharma, M. D., Smith, D. M., Burgess, R. E., Mazzella, F., Antonia, S. J., Mellor, A. L. and Munn, D. H., 2003, Pattern of recruitment of immunoregulatory antigen-presenting cells in malignant melanoma, *Lab Invest* **83**(10):1457–1466

37. Lee, K. P., Taylor, C., Petryniak, B., Turka, L. A., June, C. H. and Thompson, C. B., 1990, The genomic organization of the CD28 gene. Implications for the regulation of CD28 mRNA expression and heterogeneity, *J Immunol* **145**(1):344–352

38. Lindstein, T., June, C. H., Ledbetter, J. A., Stella, G. and Thompson, C. B., 1989, Regulation of lymphokine messenger RNA stability by a surface-mediated T cell activation pathway, *Science* **244**(4902):339–343

39. Linsley, P. S., Brady, W., Urnes, M., Grosmaire, L. S., Damle, N. K. and Ledbetter, J. A., 1991, CTLA-4 is a second receptor for the B cell activation antigen B7, *J Exp Med* **174**(3):561–569

40. Lokhorst, H. M., Lamme, T., de Smet, M., Klein, S., de Weger, R. A., van Oers, R. and Bloem, A. C., 1994, Primary tumor cells of myeloma patients induce interleukin-6 secretion in long-term bone marrow cultures, *Blood* **84**(7):2269–2277

41. Lokhorst, H. M., Sonneveld, P. and Verdonck, L. F., 1999, Intensive treatment for multiple myeloma: where do we stand? *Br J Haematol* **106**(1):18–27

42. Martin, P. J., Ledbetter, J. A., Morishita, Y., June, C. H., Beatty, P. G. and Hansen, J. A., 1986, A 44 kilodalton cell surface homodimer regulates interleukin 2 production by activated human T lymphocytes, *J Immunol* **136**(9):3282–3287

43. Mateo, G., Castellanos, M., Rasillo, A., Gutierrez, N. C., Montalban, M. A., Martin, M. L., Hernandez, J. M., Lopez-Berges, M. C., Montejano, L., Blade, J., Mateos, M. V., Sureda, A., de la Rubia, J., Diaz-Mediavilla, J., Pandiella, A., Lahuerta, J. J., Orfao, A. and San Miguel, J. F., 2005, Genetic abnormalities and patterns of antigenic expression in multiple myeloma, *Clin Cancer Res* **11**(10):3661–3667

44. Matsuo, Y., Drexler, H. G., Harashima, A., Okochi, A., Hasegawa, A., Kojima, K. and Orita, K., 2004, Induction of CD28 on the new myeloma cell line MOLP-8 with t(11;14)(q13;q32) expressing delta/lambda type immunoglobulin, *Leukemia Res* **28**(8):869–877

45. Mellor, A. L., Chandler, P., Lee, G. K., Johnson, T., Keskin, D. B., Lee, J. and Munn, D. H., 2002, Indoleamine 2,3-dioxygenase, immunosuppression and pregnancy, *J Reprod Immunol* **57**(1–2):143–150

46. Mellor, A. L. and Munn, D. H., 1999, Tryptophan catabolism and T-cell tolerance: immunosuppression by starvation? *Immunol Today* **20**(10):469–473

47. Minges Wols, H. A., Underhill, G. H., Kansas, G. S. and Witte, P. L., 2002, The role of bone marrow-derived stromal cells in the maintenance of plasma cell longevity, *J Immunol* **169**(8):4213–4221

48. Moreaux, J., Legouffe, E., Jourdan, E., Quittet, P., Reme, T., Lugagne, C., Moine, P., Rossi, J. F., Klein, B. and Tarte, K., 2004, BAFF and APRIL protect myeloma cells from apoptosis induced by interleukin 6 deprivation and dexamethasone, *Blood* **103**(8):3148–3157

49. Munn, D. H., Sharma, M. D., Lee, J. R., Jhaver, K. G., Johnson, T. S., Keskin, D. B., Marshall, B., Chandler, P., Antonia, S. J., Burgess, R., Slingluff, C. L., Jr. and Mellor, A. L., 2002, Potential regulatory function of human dendritic cells expressing indoleamine 2,3-dioxygenase, *Science* **297**(5588):1867–1870

50. Munn, D. H., Sharma, M. D. and Mellor, A. L., 2004, Ligation of B7-1/B7-2 by human CD4+ T cells triggers indoleamine 2,3-dioxygenase activity in dendritic cells, *J Immunol* **172**(7):4100–4110

51. Orabona, C., Grohmann, U., Belladonna, M. L., Fallarino, F., Vacca, C., Bianchi, R., Bozza, S., Volpi, C., Salomon, B. L., Fioretti, M. C., Romani, L. and Puccetti, P., 2004, CD28 induces immunostimulatory signals in dendritic cells via CD80 and CD86, *Nat Immunol* **5**(11):1134–1142

52. Pellat-Deceunynck, C., Bataille, R., Robillard, N., Harousseau, J. L., Rapp, M. J., Juge-Morineau, N., Wijdenes, J. and Amiot, M., 1994, Expression of CD28 and CD40 in human myeloma cells: a comparative study with normal plasma cells, *Blood* **84**(8):2597–2603

53. Pope, B., Brown, R. D., Gibson, J., Yuen, E. and Joshua, D., 2000, B7-2-positive myeloma: incidence, clinical characteristics, prognostic significance, and implications for tumor immunotherapy, *Blood* **96**(4):1274–1279

54. Qiu, Y. H., Sun, Z. W., Shi, Q., Su, C. H., Chen, Y. J., Shi, Y. J., Tao, R., Ge, Y. and Zhang, X. G., 2005, Apoptosis of multiple myeloma cells induced by agonist monoclonal antibody against human CD28, *Cell Immunol* **236**(1–2):154–160

55. Rettig, M. B., Ma, H. J., Vescio, R. A., Pold, M., Schiller, G., Belson, D., Savage, A., Nishikubo, C., Wu, C., Fraser, J., Said, J. W. and Berenson, J. R., 1997, Kaposi's sarcoma-associated herpesvirus infection of bone marrow dendritic cells from multiple myeloma patients, *Science* **276**(5320):1851–1854

56. Robillard, N., Jego, G., Pellat-Deceunynck, C., Pineau, D., Puthier, D., Mellerin, M. P., Barille, S., Rapp, M. J., Harousseau, J. L., Amiot, M. and Bataille, R., 1998, CD28, a marker associated with tumoral expansion in multiple myeloma, *Clin Cancer Res* **4**(6):1521–1526

57. Rudd, C. E. and Raab, M., 2003, Independent CD28 signaling via VAV and SLP-76: a model for in trans costimulation, *Immunol Rev* **192**:32–41

58. Sadra, A., Cinek, T., Arellano, J. L., Shi, J., Truitt, K. E. and Imboden, J. B., 1999, Identification of tyrosine phosphorylation sites in the CD28 cytoplasmic domain and their role in the costimulation of Jurkat T cells, *J Immunol* **162**(4):1966–1973

59. Said, J. W., Rettig, M. R., Heppner, K., Vescio, R. A., Schiller, G., Ma, H. J., Belson, D., Savage, A., Shintaku, I. P., Koeffler, H. P., Asou, H., Pinkus, G., Pinkus, J., Schrage, M., Green, E. and Berenson, J. R., 1997, Localization of Kaposi's sarcoma-associated herpesvirus in bone marrow biopsy samples from patients with multiple myeloma, *Blood* **90**(11):4278–4282

60. Santiago-Schwarz, F., Tucci, J. and Carsons, S. E., 1996, Endogenously produced interleukin 6 is an accessory cytokine for dendritic cell hematopoiesis, *Stem Cells* **14**(2):225–231

61. Shahinian, A., Pfeffer, K., Lee, K. P., Kundig, T. M., Kishihara, K., Wakeham, A., Kawai, K., Ohashi, P. S., Thompson, C. B. and Mak, T. W., 1993, Differential T cell costimulatory requirements in CD28-deficient mice, *Science* **261**(5121):609–612

62. Shapiro, V. S., Mollenauer, M. N. and Weiss, A., 2001, Endogenous CD28 expressed on myeloma cells up-regulates interleukin-8 production: implications for multiple myeloma progression, *Blood* **98**(1):187–193

63. Sharpe, A. H. and Freeman, G. J., 2002, The B7-CD28 superfamily, *Nat Rev Immunol* **2**(2):116–126

64. Thompson, C. B., Lindsten, T., Ledbetter, J. A., Kunkel, S. L., Young, H. A., Emerson, S. G., Leiden, J. M. and June, C. H., 1989, CD28 activation pathway regulates the production of multiple T-cell-derived lymphokines/cytokines, *Proc Natl Acad Sci USA* **86**(4):1333–1337

65. van de Donk, N. W., Lokhorst, H. M. and Bloem, A. C., 2005, Growth factors and antiapoptotic signaling pathways in multiple myeloma, *Leukemia* **19**(12):2177–2185

66. Venuprasad, K., Parab, P., Prasad, D. V., Sharma, S., Banerjee, P. R., Deshpande, M., Mitra, D. K., Pal, S., Bhadra, R., Mitra, D. and Saha, B., 2001, Immunobiology of CD28 expression on human neutrophils. I. CD28 regulates neutrophil migration by modulating CXCR-1 expression, *Eur J Immunol* **31**(5):1536–1543

67. Waterhouse, P., Penninger, J. M., Timms, E., Wakeham, A., Shahinian, A., Lee, K. P., Thompson, C. B., Griesser, H. and Mak, T. W., 1995, Lymphoproliferative disorders with early lethality in mice deficient in Ctla-4, *Science* **270**(5238):985–988

68. Wykes, M. and MacPherson, G., 2000, Dendritic cell-B-cell interaction: dendritic cells provide B cells with CD40-independent proliferation signals and CD40-dependent survival signals, *Immunology* **100**(1):1–3

69. Zhang, X. G., Olive, D., Devos, J., Rebouissou, C., Ghiotto-Ragueneau, M., Ferlin, M. and Klein, B., 1998, Malignant plasma cell lines express a functional CD28 molecule, *Leukemia* **12**(4):610–618

70. Zhou, L. J. and Tedder, T. F., 1995, A distinct pattern of cytokine gene expression by human CD83+ blood dendritic cells, *Blood* **86**(9):3295–3301

Initial TCR Transgenic Precursor Frequency Alters Functional Behaviour of CD8 T Cells Responding to Acute Infection

Thomas Wirth and John T. Harty(✉)

The ability to mount antigen-specific T cell immune responses against invading pathogens is one of the key features of the adaptive immune system.[8] Following infection dendritic cells present antigenic peptide–MHC complexes on their surface to naïve CD8 T cells patrolling secondary lymphoid organs.[14] Interaction of these dendritic cells and antigen-specific naïve CD8 T cell precursors initiates a vigorous proliferative response resulting in multiple divisions and differentiation into an effector population. This initial expansion phase is then followed by a sharp decline in T cell numbers with 90–95% of all responding T cells succumbing to apoptotic cell death. The remaining T cells form a stable pool of memory cells capable of protecting the host against subsequent challenges with the same pathogen.[3] In order to study the kinetics of T cell expansion and the process of memory generation and maintenance, the identification of CD8 T cells specific for defined epitopes is of crucial importance. Only the continuous monitoring of antigen-specific T cells after infection or vaccination allows for the identification and characterization of T cells at different stages of their differentiation. Ideally, the identification of antigen-specific cells should encompass T cells at all stages of differentiation from naïve precursors to memory cells without manipulations that change functional behaviour.

With the introduction of intracellular cytokine-staining assays, T cell receptor (TCR) transgenic mice[18] and major histocompatibility complex (MHC) tetramers,[1] various models have become available that fulfil the requirements for reliable and reproducible identification of antigen-specific CD8 T cells. For intracellular cytokine-staining assays CD8 T cells have to be stimulated ex vivo with peptide-pulsed cells, thus limiting the assays' application in vivo. Similarly, the use of MHC tetramers requires binding of the tetramers to their corresponding TCR to identify the antigen-specific T cell subset. In contrast, TCR transgenic T cells

John T. Harty

Department of Microbiology, Interdisciplinary Graduate Program in Immunology,
University of Iowa, 51 Newton Road, 3-512 Bowen Science Building,
Iowa City, IA 52242,USA

S.P. Schoenberger et al. (eds.) *Crossroads between Innate and Adaptive Immunity II*,
doi: 10.1007/978-0-387-79311-5_7, © Springer Science + Business Media, LLC 2009

require no ex vivo manipulation for identification as virtually all the CD8 T cells of a transgenic animal carry the TCR of the desired antigen specificity. Due to this specificity, adoptive transfer experiments can be employed to directly compare TCR transgenic T cells with different genetic characteristics in the same host. However, to study these transgenic T cells in the context of a CD8 T cell pool with a diverse TCR repertoire, adoptive transfer experiments have to be performed where varying numbers of the TCR transgenic T cells are transferred to a naïve host before a challenge with the corresponding antigen. A Pubmed search for studies using TCR transgenic T cells currently yields more than 3,000 publications that have deployed the strategy of adoptive transfers to study adaptive immune responses. Following the setup of the initial studies with TCR transgenic T cells, most of the published experiments are based on the transfer of relatively large numbers ($>10^6$) of T cells to facilitate identification of the transferred T cell population prior to immunization. Although rarely addressed, these adoptive transfer experiments were believed to closely reflect the situation observed in an endogenous immune response to infection or immunization. However, subsequent studies attempting to quantify the number of endogenous naïve CD8 T cell precursors revealed that the natural TCR repertoire contains approximately 10–1,000 CD8 T cells specific for a single epitope.[5,6,9,19,25,26] Thus most of the studies to date are performed with TCR transgenic T cells that outnumber endogenous naïve precursors by several orders of magnitude. The consequences of this increased abundance of naïve CD8 T cells for their interaction with limited numbers of matured dendritic cells and the proliferation of the primed CD8 T cells remain largely unknown.

With increasing awareness that adoptive transfer experiments should try to mimic endogenous immune responses as accurately as possible, a small number of studies have recently tried to address the importance of initial T cell frequency in subsequent responses to antigen challenge.[10,13,20,22] The results of these studies suggest that critical aspects of both CD4 and CD8 T cell behaviour depend on the numbers of naïve precursors present at the time of infection. In CD4 T cells, survival of both naïve precursors and memory CD4 T cells was critically dependent on the number of naïve TCR transgenic T cells transferred with preferential survival of clones at a low, more physiological frequency.[13] In addition, the number of divisions during the expansion phase was inversely correlated with the number of transferred precursors.[11,13] As a consequence, early reports claiming a low proliferative potential for CD4 T cells[12] have recently been questioned by results from experiments with low numbers of adoptively transferred TCR transgenic precursors that demonstrate a similar magnitude of clonal expansion for CD4 and CD8 T cells in acute infections.[11]

Similar to the results for CD4 T cells, CD8 T cell characteristics change dramatically with increased precursor frequency. The spectrum of affected characteristics includes crucial components of adaptive immune responses like T cell kinetics and proliferation, surface-marker expression, effector function and memory T cell generation. The nature of these changes and their importance for a more physiological design of immune response studies shall be matter of this review.

1 CD8 T Cell Proliferation and Kinetics

In acute infections CD8 T cells need to undergo extensive proliferation to generate effector cell numbers sufficient to protect a host from an invading pathogen.[8] With regard to the maximum number of effector cells generated from different numbers of TCR transgenic CD8 T cells, a ceiling effect has been proposed, based on experiments performed with peptide-pulsed dendritic cell immunization such that different numbers of input TCR transgenic cells (OT-I) generate the same number of effector cells at day seven after immunization.[20] While this may be true for DC immunizations, studies from our laboratory show that after *Listeria monocytogenes* infection higher numbers of transferred precursors generally generate higher numbers of CD8 T cells at the peak of expansion.[2] This observation might suggest that when transferred in large numbers T cells undergo more vigorous clonal expansion. However, naïve precursors transferred in low numbers generated substantially more effector cells on a per-cell basis indicating that these cells had undergone additional rounds of division. Direct comparison of two groups with 5×10^1 or 5×10^5 TCR transgenic T cells revealed an approximately 1,000× increase in effector cells created on a per-cell basis in the group with lower input numbers. Different input numbers not only changed the observed peak numbers and the number of individual T cell divisions but also altered the overall kinetics of the immune response (Fig. 1a). While lower numbers of TCR transgenic cells preserved the peak at day seven as seen in the endogenous response to LM-OVA infection, higher input numbers shifted the peak of the response to time points as early as day 5. Notably, in the groups with the highest number of T cells transferred, more than 50% of the ensuing contraction had already occurred on day 7. In adoptive transfer experiments, these results must be taken into consideration for the calculation of contraction percentages which requires a precise identification of peak numbers. One way to ensure an accurate analysis involves taking individual, daily blood samples to identify peak, contraction phase and memory level in each of the animals studied.

The fact that the number of transferred CD8 T cells affects the magnitude of proliferation may hold the key to the explanation of other differences observed in T cell behaviour. Amongst others, phenotype and memory lineage commitment may be determined by initial activation levels and the proliferative history of T cells. Thus, the number of divisions T cells go through determines many of the functional characteristics these cells show at later stages of their differentiation.

2 Alteration of T Cell Phenotype

To assess the quality of individual T cells in an immune response a plethora of phenotypic markers have been described that allow for the identification of permanent or transient subsets of T cells at multiple stages of their differentiation to memory. Some of these markers have been used to identify early memory precursors,

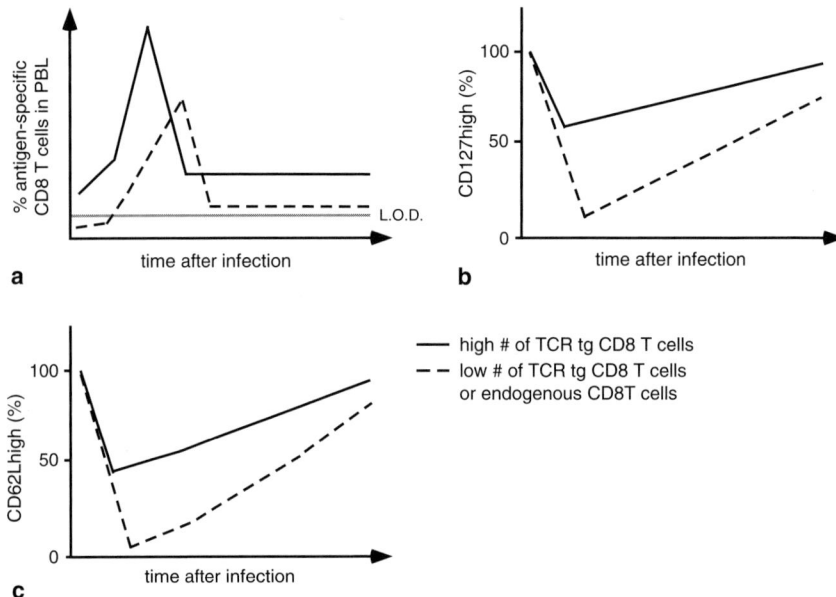

Fig. 1 Input numbers of TCR transgenic cells alter T cell kinetics, phenotype and memory line-age commitment. (**a**) In response to acute infection, high numbers of adoptively transferred TCR transgenic T cells alter the kinetics of the immune response resulting in earlier peak of expansion, earlier contraction and higher levels of memory. (**b**) Expression of CD127 depends on the transferred precursor frequency. Higher numbers of transferred precursors lead to reduced CD127 downregulation and faster transition to a CD127high phenotype. (**c**) High numbers of adoptively transferred T cells show reduced downregulation of CD62L compared to low numbers of transferred cells or endogenous immune responses. Higher expression of CD62L in groups with large numbers of transferred precursors thus leads to a preferential central memory phenotype

while others play a role in the identification of different memory subsets.[17,22] While it is widely accepted that expression of these markers may vary depending on the infectious agent or the inflammatory milieu the expression pattern of these markers is rather consistent when the same pathogen is used in every experiment. As an example, T cells expressing the IL-7 receptor alpha unit (CD127) have been suggested to identify effector T cells at the peak of expansion that preferentially become memory cells due to the expression of anti-apoptotic Bcl-2 protein.[15,17] When expression of CD127 was assessed at the peak of expansion in groups with different input numbers of T cells in our laboratory, only the groups with low precursor frequency showed the typical CD127lo phenotype observed in endogenous immune responses (Fig. 1b). In contrast, T cells derived from larger numbers of precursors showed a predominant CD127hi phenotype with 70% of all T cells expressing CD127 as early as day 5. The expression of CD127 occurred despite an overall similar percentage of contraction in all groups. In conclusion, large numbers of T cells not only alter the expression pattern of CD127 but also prevent the use of this marker to identify memory precursor effector cells.

The differences in phenotype between groups with different input numbers of TCR transgenic T cells were not limited to CD127. While low input yielded the typical effector phenotype (CD62Llo, CD127lo, KLRGhi, CD43hi), increasing numbers of T cell precursors resulted in an intermediate phenotype that closely resembled the expression pattern of naïve/memory T cells. Again, these results suggest that increasing the abundance of naïve precursors might affect the activation status of the T cells leading to an incomplete activation phenotype or accelerated progression to memory.

3 Memory CD8 T Cell Lineage Commitment

In general, memory CD8 T cells can be divided into effector memory and central memory T cells based on their expression of CD62L and CCR7.[28] While central memory T cells express CD62L and CCR7 at high levels which enables them to enter secondary lymphoid organs, effector memory T cells retain the CD62Llo CCR7lo expression pattern which excludes them from lymph nodes and directs them to tertiary tissues.[23] Due to their differential localization in the host a division of labour has been proposed for the two subsets with effector memory cells exerting immediate effector function in tertiary tissues and central memory cells as a reserve population capable of vigorous proliferation.[27,30] The relationship between these two populations and the question whether they represent mutually exclusive lineages of memory T cells has been subject of numerous publications with often contradictory results. Evidence for possible conversion of central memory to effector memory cells comes in part from in vitro studies showing a gradual loss of CCR7 over time.[21,28] In contrast, detailed analysis of the TCR repertoire of the two populations in human blood has suggested that central memory and effector memory T cells represent separate lineages.[4] To further complicate the matter, convincing evidence comes from in vivo studies that effector memory T cells can convert to central memory T cells as well.[30]

In a recent study Marzo et al.[22] addressed the question of memory lineage commitment in groups of mice with varying numbers of adoptively transferred or endogenous naïve precursors. The results show conclusively that with increasing input numbers of naïve T cells effector memory T cells gain the ability to rapidly convert to central memory T cells (Fig. 1c). In contrast, the two memory subsets showed only minimal signs of interconversion in groups with low numbers of transferred precursors or in endogenous polyclonal immune responses. As a possible caveat of this experiment the adoptively transferred T cell populations were only monitored for 30 days which could be insufficient to detect slow conversion of the memory T cell subsets. Results from our own laboratory show that effector cells derived from high numbers of precursors re-acquired CD62L expression substantially faster than those derived from low numbers.[2] In some high input number groups, 50% of all effector T cells stained positive for CD62L as early as day five after infection, and these groups maintained a higher percentage of CD62L+ memory

CD8 T cells compared to groups with physiological numbers of input precursors. These results indicate that the ratio of central memory vs. effector memory cells generated in an immune response is not fixed but subject to change depending on the number of input cells. Since the two memory subsets differ in their ability to defend the host against re-infection the protective capacity of vaccines using high numbers of adoptively transferred TCR transgenic T cells might be misinterpreted.

4 Influence of Adoptively Transferred T Cells on the Endogenous Immune Response

The ultimate goal of every infection study is to mimic 'natural' immune responses as closely as possible. In endogenous immune responses T cells specific for one epitope represent only a small fraction of all the T cells responding to an invading pathogen. Due to their low numbers these naïve precursors are not likely to face excessive competition for antigen or other signals presented by matured dendritic cells. However, with increasing numbers of naïve precursors both the quantity and the quality of DC–T cell interactions could be altered. As a result, either the magnitude of stimulation or the number of T cells recruited into the response (or both) may vary considerably. When adoptive transfer experiments were performed in our laboratory with different numbers of TCR transgenic OT-I cells, an inverse correlation between the numbers of transferred OT-I cells and the magnitude of the endogenous immune response became evident (Fig. 2). In fact, even intermediate numbers (5×10^3 OT-I cells) led to nearly undetectable numbers of endogenous CD8 T cells.

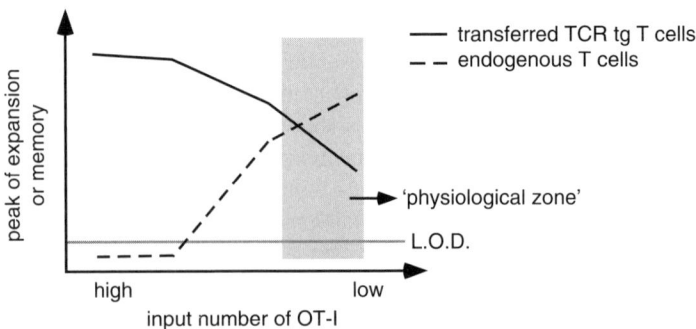

Fig. 2 Input numbers of TCR transgenic T cells affect the endogenous immune response. With increasing numbers of adoptively transferred TCR transgenic precursors the endogenous immune response to the same antigen is suppressed. Lowering the number of transferred T cells restores the endogenous immune response while preserving the expansion of the transferred TCR transgenic population

Why is the preservation of the endogenous immune response important in adoptive transfer experiments? As determined in our laboratory for TCR transgenic OT-I and P14 cells, the number of transferred cells needed to suppress the endogenous response depends on the individual system. The presence of an endogenous response indicates that the number of transferred cells is similar to that of the endogenous naïve precursors. As further proof of this assumption the phenotype of endogenous and exogenous T cells can be compared in the same host to verify that the predictive power of surface phenotype is not compromised. While comparison of TCR transgenic and endogenous T cells may not be feasible for every single experiment, these data strongly suggest that careful titration experiments should be performed at least once for every TCR transgenic system used.

5 Conclusions

Experiments with TCR transgenic T cells have made valuable contributions to the field of immunological research and provided new insights into the behaviour of epitope-specific subsets of T cells in infection. Despite the popularity of adoptive transfer systems only few studies have performed systematic comparisons between adoptive transfer experiments and endogenous immune responses, and results obtained from studies using large numbers of adoptively transferred TCR transgenic T cells have not been questioned.

The few studies discussed in this review, however, show unanimously that the number of transferred TCR transgenic T cells alters critical aspects of an immune response like proliferation, kinetics, phenotype and memory generation. With increasing abundance of naïve TCR transgenic precursors, T cells undergo less rounds of division, reach higher peak numbers at earlier time points, show an altered phenotype and become preferentially central memory T cells. As a consequence, performing adoptive transfer experiments without careful prior titration experiments can for example lead to misinterpretation of the magnitude of expansion and contraction or the quality of the T cells primed during the response.

The exact reasons for the modified behaviour of the T cells remain unknown but are most likely to be found in the initial encounter of T cells and dendritic cells where higher numbers of naïve precursors compete for antigen and possibly other signals presented by dendritic cells. Competition for dendritic cells could lead to reduced levels of T cell activation thus leaving a different 'imprint' on these cells during their subsequent proliferation.[16] This assumption may have far-reaching consequences since most of the multi-photon studies used to illustrate interactions of dendritic cell with naïve T cells have used large numbers of adoptively transferred precursors.[7,24,29] Most of the studies describe at least three distinct phases of interaction with the first phase displaying increased mobility of the T cells within a lymph node for a few hours. In the second phase T cells form stable contacts with APCs followed by again increased mobility and more transient DC–T cell interactions before egress from the lymph node. Whether this classification and the duration of each individual phase are representative of endogenous immune responses where dendritic cells encounter thousand-fold less naïve T cell precursors remains to be determined.

The results of the studies clearly argue that adoptive transfer experiments may not mimic endogenous immune responses adequately. Yet the same experiments demonstrate that when low, more physiological numbers of T cells are transferred, the induced immune response was virtually identical to the endogenous response. When transfer numbers approximated the expected natural precursor frequency the transferred TCR transgenic T cells became part of a robust endogenous response which might represent the most physiological situation achievable that still allows for the identification of antigen-specific CD8 T cells. As a part of this endogenous response, the transferred T cells showed the same phenotype expression pattern as the endogenous T cells for all the phenotypic markers examined which strongly supports their use to interpret T cell quality and function. Even with numbers as low as 5×10^1 transferred TCR transgenic naïve precursors the immune response remained readily detectable in all tissues at effector, contraction and memory phase, indicating that the readout of the experiments was not compromised. The number of TCR transgenic T cells needed for such experiments can easily be obtained from blood samples which represents a major advantage when the number of donor mice is limited.

In summary, the current state of knowledge argues against the adoptive transfer of large numbers of TCR transgenic T cells as a natural, physiological model for studying immune responses. To combine the feasibility of adoptive transfers with reliable and reproducible results, a more physiological setup with lower numbers of precursors is of critical importance.

References

1. Altman, J.D., Moss, P.A., Goulder, P.J., Barouch, D.H., McHeyzer-Williams, M.G., Bell, J.I., McMichael, A.J., and Davis, M.M. (1996). Phenotypic analysis of antigen-specific T lymphocytes. *Science* **274,** 94–96
2. Badovinac, V.P., Haring, J.S., and Harty, J.T. (2007). Initial T cell receptor transgenic cell precursor frequency dictates critical aspects of the CD8(+) T cell response to infection. *Immunity* **26,** 827–841
3. Badovinac, V.P., and Harty, J.T. (2006). Programming, demarcating, and manipulating CD8+ T-cell memory. *Immunological Reviews* **211,** 67–80
4. Baron, V., Bouneaud, C., Cumano, A., Lim, A., Arstila, T.P., Kourilsky, P., Ferradini, L., and Pannetier, C. (2003). The repertoires of circulating human CD8(+) central and effector memory T cell subsets are largely distinct. *Immunity* **18,** 193–204
5. Blattman, J.N., Antia, R., Sourdive, D.J., Wang, X., Kaech, S.M., Murali-Krishna, K., Altman, J.D., and Ahmed, R. (2002). Estimating the precursor frequency of naive antigen-specific CD8 T cells. *Journal of Experimental Medicine* **195,** 657–664
6. Bousso, P., Casrouge, A., Altman, J.D., Haury, M., Kanellopoulos, J., Abastado, J.P., and Kourilsky, P. (1998). Individual variations in the murine T cell response to a specific peptide reflect variability in naive repertoires. *Immunity* **9,** 169–178

7. Bousso, P., and Robey, E. (2003). Dynamics of CD8+ T cell priming by dendritic cells in intact lymph nodes. *Nature Immunology* **4,** 579–585
8. Butz, E.A., and Bevan, M.J. (1998). Massive expansion of antigen-specific CD8+ T cells during an acute virus infection. *Immunity* **8,** 167–175
9. Casrouge, A., Beaudoing, E., Dalle, S., Pannetier, C., Kanellopoulos, J., and Kourilsky, P. (2000). Size estimate of the alpha beta TCR repertoire of naive mouse splenocytes. *Journal of Immunology* **164,** 5782–5787
10. Ford, M.L., Koehn, B.H., Wagener, M.E., Jiang, W., Gangappa, S., Pearson, T.C., and Larsen, C.P. (2007). Antigen-specific precursor frequency impacts T cell proliferation, differentiation, and requirement for costimulation. *Journal of Experimental Medicine* **204,** 299–309
11. Foulds, K.E., and Shen, H. (2006). Clonal competition inhibits the proliferation and differentiation of adoptively transferred TCR transgenic CD4 T cells in response to infection. *Journal of Immunology* **176,** 3037–3043
12. Foulds, K.E., Zenewicz, L.A., Shedlock, D.J., Jiang, J., Troy, A.E., and Shen, H. (2002). Cutting edge: CD4 and CD8 T cells are intrinsically different in their proliferative responses. *Journal of Immunology* **168,** 1528–1532
13. Hataye, J., Moon, J.J., Khoruts, A., Reilly, C., and Jenkins, M.K. (2006). Naive and memory CD4+ T cell survival controlled by clonal abundance. *Science* **312,** 114–116
14. Heath, W.R., Belz, G.T., Behrens, G.M., Smith, C.M., Forehan, S.P., Parish, I.A., Davey, G.M., Wilson, N.S., Carbone, F.R., and Villadangos, J.A. (2004). Cross-presentation, dendritic cell subsets, and the generation of immunity to cellular antigens. *Immunological Reviews* **199,** 9–26
15. Huster, K.M., Busch, V., Schiemann, M., Linkemann, K., Kerksiek, K.M., Wagner, H., and Busch, D.H. (2004). Selective expression of IL-7 receptor on memory T cells identifies early CD40L-dependent generation of distinct CD8+ memory T cell subsets. *Proceedings of the National Academy of Sciences of the United States of America* **101,** 5610–5615
16. Kaech, S.M., and Ahmed, R. (2001). Memory CD8+ T cell differentiation: initial antigen encounter triggers a developmental program in naive cells. *Nature Immunology* **2,** 415–422
17. Kaech, S.M., Tan, J.T., Wherry, E.J., Konieczny, B.T., Surh, C.D., and Ahmed, R. (2003). Selective expression of the interleukin 7 receptor identifies effector CD8 T cells that give rise to long-lived memory cells. *Nature Immunology* **4,** 1191–1198
18. Kearney, E.R., Pape, K.A., Loh, D.Y., and Jenkins, M.K. (1994). Visualization of peptide-specific T cell immunity and peripheral tolerance induction in vivo. *Immunity* **1,** 327–339
19. Kedzierska, K., Day, E.B., Pi, J., Heard, S.B., Doherty, P.C., Turner, S.J., and Perlman, S. (2006). Quantification of repertoire diversity of influenza-specific epitopes with predominant public or private TCR usage. *Journal of Immunology* **177,** 6705–6712
20. Kemp, R.A., Powell, T.J., Dwyer, D.W., and Dutton, R.W. (2004). Cutting edge: regulation of CD8+ T cell effector population size. *Journal of Immunology* **173,** 2923–2927
21. Lanzavecchia, A., and Sallusto, F. (2000). Dynamics of T lymphocyte responses: intermediates, effectors, and memory cells. *Science* **290,** 92–97
22. Marzo, A.L., Klonowski, K.D., Le Bon, A., Borrow, P., Tough, D.F., and Lefrancois, L. (2005). Initial T cell frequency dictates memory CD8+ T cell lineage commitment. *Nature Immunology* **6,** 793–799
23. Masopust, D., Vezys, V., Marzo, A.L., and Lefrancois, L. (2001). Preferential localization of effector memory cells in nonlymphoid tissue. *Science* **291,** 2413–2417
24. Mempel, T.R., Henrickson, S.E., and Von Andrian, U.H. (2004). T-cell priming by dendritic cells in lymph nodes occurs in three distinct phases. *Nature* **427,** 154–159
25. Moon, J.J., Chu, H.H., Pepper, M., McSorley, S.J., Jameson, S.C., Kedl, R.M., and Jenkins, M.K. (2007). Naive CD4(+) T cell frequency varies for different epitopes and predicts repertoire diversity and response magnitude. *Immunity* **27,** 203–213
26. Pewe, L.L., Netland, J.M., Heard, S.B., and Perlman, S. (2004). Very diverse CD8 T cell clonotypic responses after virus infections. *Journal of Immunology* **172,** 3151–3156
27. Reiner, S.L., Sallusto, F., and Lanzavecchia, A. (2007). Division of labor with a workforce of one: challenges in specifying effector and memory T cell fate. *Science* **317,** 622–625

28. Sallusto, F., Lenig, D., Forster, R., Lipp, M., and Lanzavecchia, A. (1999). Two subsets of memory T lymphocytes with distinct homing potentials and effector functions. *Nature* **401,** 708–712

29. Stoll, S., Delon, J., Brotz, T.M., and Germain, R.N. (2002). Dynamic imaging of T cell–dendritic cell interactions in lymph nodes. *Science* **296,** 1873–1876

30. Wherry, E.J., Teichgraber, V., Becker, T.C., Masopust, D., Kaech, S.M., Antia, R., von Andrian, U.H., and Ahmed, R. (2003). Lineage relationship and protective immunity of memory CD8 T cell subsets. *Nature Immunology* **4,** 225–234

Effects of Cancer Immunotherapy Regimens on Primary vs. Secondary Immune Responses and the Potential Impact on Long-Term Antitumor Responses

Kory L. Alderson and William J. Murphy(✉)

1 Memory T Cell Responses and Cancer

While the use of immunotherapy to promote T cell responses for the treatment of certain cancers has made significant progress, it has been hampered with relatively low response rates with regard to overall survival and has been applied to a limited number of cancers. One area that is lacking thus far is an understanding of the importance (or not) of memory T cell generation after immunotherapy in cancer. There are multiple models of memory T cell formation.[1,2] When properly primed, memory T cells are tenacious and capable effectors that can remove antigen-positive cells before any signs of re-encounter have occurred (Fig. 1). For this reason, the generation of memory T cells in an individual with cancer is desirable for durable and sustained antitumor responses.[1,2] The generation of antigen-specific T cell responses to cancer has been the focus of some groups attempting to increase the numbers of antigen-specific T cells in patients through ex vivo expansion followed by adoptive transfer.[3] However the reported increase is in complete response rates without discussion of the overall survival time of patients. This may be an indicator that immunological memory is not being attained or that downregulation pathways may be exerted after immunotherapy.[4–6] For adoptive lymphocyte transfer to be efficacious, it is imperative to understand how T cells persist in an environment where extensive immune stimulation is being applied. Our laboratory has recently introduced the question: what effect does the use of potent immune stimuli have on immunological memory? Our previous work demonstrated that immunotherapy resulting in high production of IFNγ impaired secondary T cell responses, despite initial antitumor responses capable of eliciting tumor regression.[7,8] These findings have created a paradox with respect to the use of strong immunotherapeutic agents to treat cancer; while strong immune stimulation may be necessary for the immune system to adequately recognize and destroy weakly immunogenic tumors, their use may also be impairing the development of immunological memory or sustained antitumor responses.

William J. Murphy
Department of Microbiology and Immunology, University of Nevada Reno
School of Medicine, Reno, NV 89557, USA

S.P. Schoenberger et al. (eds.) *Crossroads between Innate and Adaptive Immunity II,*
doi: 10.1007/978-0-387-79311-5_8, © Springer Science+Business Media, LLC 2009

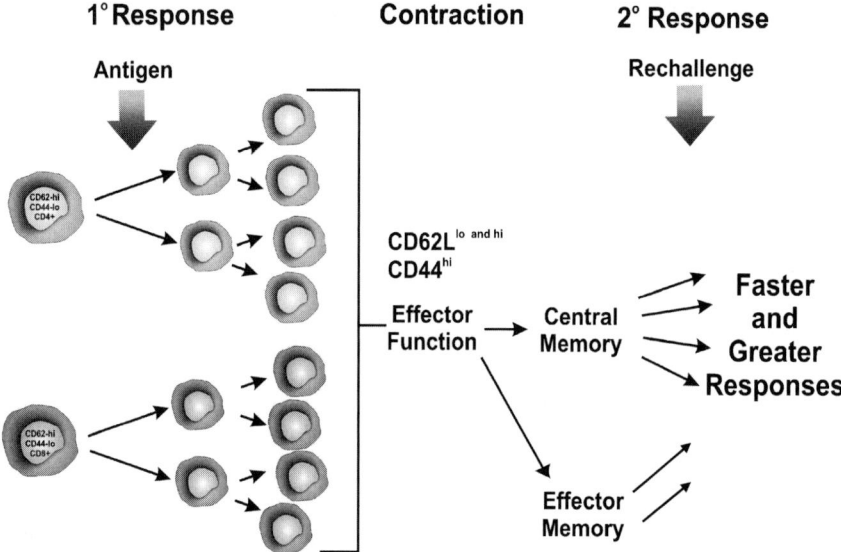

Fig. 1 Generation of a functional memory response. There are three phases to an immune response. First is a primary response where upon initial antigen encounter, antigen-specific T cells undergo rapid proliferation. Second is the contraction phase. After antigen is cleared, the expanded effector cells undergo apoptosis and few antigen-specific cells are retained as long-lasting memory cells. Finally, upon antigen re-encounter is a secondary response phase, where long-lasting memory cells are capable of responding to antigen more quickly than in the primary response

Strong cytokine stimuli has been demonstrated as having deleterious effects on antigen-specific T cells by both direct and indirect effects.[9–13] More recently, it was demonstrated that the use of potent immune stimulatory agents, such as an agonist CD40 mAb (anti-CD40) and interleukin-2 (IL-2) or CpG and IL-12, as a source of therapy can result in deletion of cells that are vital to the antigen-specific response and memory cell generation.[7,14] CD40 is an important immunological marker vital to immune activation.[15] Bartholdy et al.[14] reported a loss of LCMV-virus-specific CD8+ T cells after treatment with an anti-CD40. Their observations highlighted that despite artificial "help" to CD8+ T cells during a viral infection through anti-CD40, CD8+ T cells instead of being primed were in fact deleted. This resulted in a lack of sustained antiviral responses. Our laboratory has previously shown that combined immunotherapy with anti-CD40 and IL-2 could markedly increase the survival of tumor-bearing mice[8] (Fig. 2a). However, if mice were vaccinated with irradiated tumor prior to immunotherapeutic administration, they did not generate a significant memory response and were not protected against a later live tumor challenge[7] (Fig. 2b). Furthermore, we found that while CD8+ T cells expand following immunotherapy, CD4+ T cell numbers did not increase. Importantly, this lack of CD4+ T cell expansion was dependent on interferon gamma (IFNγ) and treatment of mice that lack IFNγ signaling, (IFNγ-deficient mice (IFNγ–/–) or IFNγ receptor-deficient (IFNγR–/–) mice) resulted in increases in CD4+ T cell numbers similar to the observed expansion in CD8+ T cells.[7]

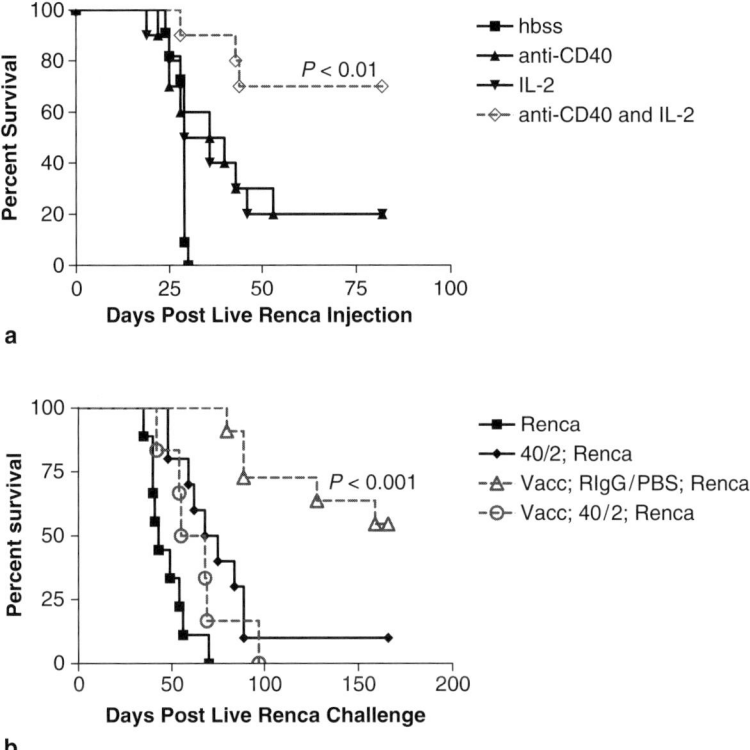

Fig. 2 Anti-CD40 and IL-2 leads to a successful initial antitumor response which does not result in a functional secondary response to tumor. (**a**) As reported previously[8] combination immunotherapy with anti-CD40 and IL-2 resulted in synergistic antitumor responses. Administration of the combination (*open diamonds*) significantly enhanced the survival time of tumor-bearing mice over the use of either agent alone (*closed triangles*). (**b**) Anti-CD40 and IL-2 did not support the generation of functional memory to tumor antigen.[7] Administration of anti-CD40 and IL-2 after vaccine challenge with irradiated tumor (*open circles*) negated protection from later live tumor challenge. However when vaccine was not followed by anti-CD40 and IL-2, mice retained functional immunological memory to tumor antigens (*open triangles*)

2 The Role of Programmed Death-1 and Programmed Death Ligand-1

Following anti-CD40 and IL-2 immunotherapy, CD8[+] T cells significantly expand in numbers, whereas CD4[+] T cell numbers remain relatively the same. This was found to be due to elevated apoptosis and necrosis of the activated CD4[+] T cells.[7] Further investigation into the mechanism underlying this impairment of CD4[+] T cell expansion revealed that cell surface expression of Fas or DR5 (as well as FasL or TRAIL) was not elevated but expression of programmed death-1 (PD-1) was elevated on the surface of CD4[+] but not CD8[+] T cells following immunotherapy (Alderson et al., submitted). PD-1 is found on most cells of hematopoietic origin and its gene expression has been associated with programmed cell death of thymocytes after TCR ligation.[16] The presence of surface PD-1 on activated T cells is important for peripheral tolerance of CD8[+] T cells to self-expressed antigens in the tissues.[17] Ligation of PD-1 by B7-H1 is also important for tolerance to self-antigens early in T cell development.[18] PD-1 expression on T cells may be used to indicate disease progression in a variety of disease settings including HIV, Hodgkin's lymphoma, and schistosomiasis.[19–21] In an autoimmune setting, such as in the case of rheumatoid arthritis, PD-1[+] CD4[+] T cells can accumulate in the synovial fluid of affected patients and represent a population of anergic T cells.[22]

Two ligands are known to recognize and bind PD-1: B7-H1 (PDL-1, CD274) and B7-DC (PDL-2, CD273).[23] The expression of B7-H1 is dependent on IFNγ and after ligation of PD-1 is described to elicit either apoptosis or senescence in the PD-1-bearing cell. B7-DC, which is primarily found on dendritic cells, is not dependent on IFNγ. B7-H1 is found on many cell types, both hematopoietic and nonhematopoietic in origin.[24] It has been shown that surface B7-H1 on tumor cells can contribute to tumor evasion of immune response as it is upregulated both in vivo and in vitro in response to IFNγ.[25] PD-1 is involved with peripheral CD8[+] T cell tolerance to self-antigens[17] and PD-1 engagement by B7-H1 has been shown to have potent inhibitory effects on immune stimulation.[26,27] Because of the potential to modulate immune responses in a positive or negative manner, the PD-1/B7-H1 pathway is under investigation with respect to cancer therapy.[28–30]

Our laboratory recently found PD-1 to be upregulated on the surface of CD4[+] T cells after treatment with anti-CD40 and IL-2 (submitted). However, when we evaluated the relative expression pattern of PD-1 on the surface of CD4[+] T cell subsets, we uncovered a differential response to immunotherapy. PD-1 is selectively upregulated on conventional CD4[+] T cells (Tconv) and not on the inhibitory class of CD4[+] T cells (Treg) expressing forkhead transcription factor-3 (Foxp3). This differential expression pattern after anti-CD40 and IL-2 allowed Treg cells to continue to expand after immunotherapy while Tconv cells did not. This results in a preferential Treg cell expansion following immunotherapy, and we found Treg cells to be as high as 40% of the CD4[+] T cell repertoire (submitted). PD-1 is similarly upregulated in both wild-type and IFNγ–/– or IFNγR–/– mice in response to immunotherapy. However, the expression of the ligand for PD-1, B7-H1, is dependent on IFNγ and is not upregulated following anti-CD40 and IL-2 in either IFNγ–/– or IFNγR–/– mice. Therefore, while the differential upregulation of PD-1 after immunotherapy on Tconv vs. Treg cells correlated with CD4[+] T cell nonexpansion, the IFNγ-dependent upregulation of B7-H1 allows for enhanced signaling through PD-1. After administration of immunotherapy, combined upregulation of PD-1 and B7-H1 had an overall dampening effect on Tconv cell expansion capabilities, but not on Treg cell expansion. This conversion within the CD4[+] T cell compartment to more largely inhibitory may be a contributing factor to the lack of memory T cell priming during the initial immune response. Treg cells have been shown to exert dampening effects on the primary responses of both CD4[+] and CD8[+] T cells and NK cells.[31,32] Therefore, while

Treg cell expansion after anti-CD40 and IL-2 administration does not appear to be hampering the initial antitumor effect, removal of Treg cells in conjunction with immunotherapy may reduce the amount of necessary drug administration with similar results in tumor regression.

The role of PD-1 on the surface of CD4+ T cell subsets has not been extensively examined. In *Helicobacter pylori* infection, PD-1 ligation by B7-H1 at the site of infection results in T cell anergy. Furthermore, coculture of CD4+ T cells from *H. pylori*-infected donors with infected epithelial cells in vitro results in an expansion of Treg cells when B7-H1 is present. Blocking B7-H1 with blocking antibodies in this model takes away the ability of Treg cells to expand.[33] Differential expression pattern of PD-1 between resting Treg cells and activated Tconv cells has been recently discussed after CD3 stimulation in vitro. This report showed that while Treg cells lack surface expression of PD-1 on their surface they do have intracellular PD-1. More importantly, CD3 activation of Tconv cells in vitro resulted in differential expression of PD-1 and CD25 which was dependent on the presence of Treg cells. In the presence of Treg cells, stimulated Tconv cells only upregulated PD-1 in response to CD3 stimulation; however, if Treg cells were not present in culture, both PD-1 and CD25 were upregulated on the surface of Tconv cells.[34] In an immunotherapy setting such as anti-CD40 and IL-2 these data suggest that removal of Treg cells in conjunction with immunotherapy may result in a more efficient ability of Tconv cells to respond to IL-2 by allowing for enhanced CD25 expression.

3   The Consequences of Immunotherapy

Immunotherapy represents powerful potential as a means to treat cancer. However, the topics discussed in this review have suggested that the application of strong stimuli as immunotherapeutic agents may be strongly opposed by regulatory mechanisms within the immune repertoire. This counteractive immunological response may have important consequences with respect to the efficacy of immunotherapy. First, a reduction in the CD4:CD8 balance coinciding with an increase in Treg cells making up the CD4+ compartment may be detrimental to the generation of sustained antitumor responses. Also, this shift in the immune repertoire, favoring cells of inhibitory phenotype, may increase the amount of drug required to gain the desired proinflammatory result. This is especially important for drugs which have proven to be effective but restricted due to toxicity, such as IL-2.[35] These latest findings regarding PD-1 and B7-H1 upregulation in response to immunotherapy have brought to light new questions surrounding the consequences of immunotherapy on the maintenance of antitumor immune responses (Fig. 3).

Expression patterns of PD-1 and B7-H1 are important indicators of disease outcome.[19–21] However, they may also be important indicators of the efficacy of immunotherapy. B7-H1 has been described as a "molecular shield" used by tumors as it is upregulated in response to IFNγ.[25] The selective upregulation of PD-1 on the surface of Tconv cells and not Treg cells after immunotherapy augments the

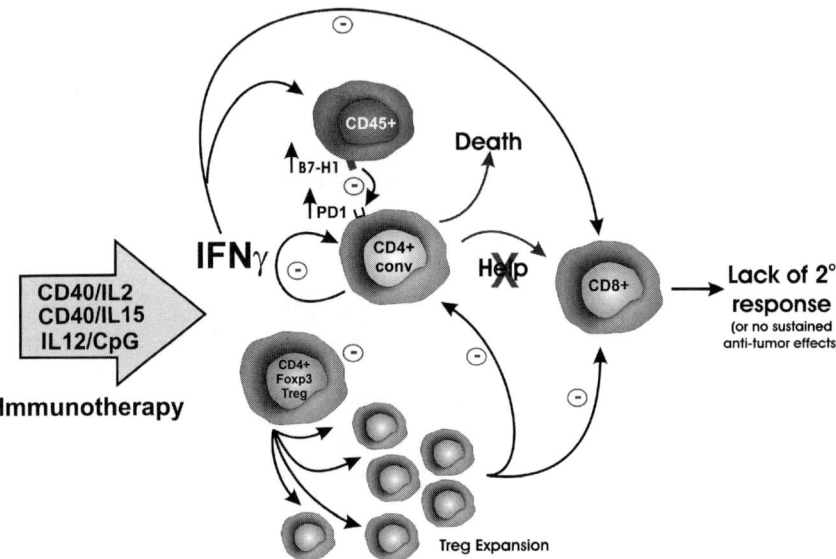

Fig. 3 Consequences of immunotherapy in the maintenance of antitumor responses. Administration of a potent stimulatory immunotherapeutic agent can result in positive stimulus to Treg cells and concurrent IFNγ-dependent negative stimulus to CD4+ Tconv cells. This occurs because of immunotherapy-induced upregulation of PD-1 on the surface of Tconv cells and not Treg cells coinciding with IFNγ-induced upregulation of B7-H1 on all hematopoietic cells. As a result, a functional secondary response is not maintained and mice are not protected against a live tumor challenge

importance of this immunological phenomenon as it highlights that after immunotherapy, both the tumor and the cells of the immune system are actively attempting to dampen the inflammatory response. Effective immunotherapeutic applications to treat cancer are sometimes associated with the development of autoimmune disease.[36] Therefore, blocking inhibitory receptors such as PD-1, B7-H1 and CTLA-4 are used as means to break tolerance to self-antigens.[18,36,37] In combination with immunotherapeutic stimuli, blockade of PD-1 and/or B7-H1 may be necessary to achieve a maximal response.

Using blocking antibodies against PD-1 and B7-H1 is best applied when used in combination. It was recently shown that combined blockade, but not either antibody alone, was capable of enhancing antitumor effects.[37] Blockade of one side of this pathway is usually met with only partial effects, and this demonstrates a problem when attempting to block signaling after immunotherapy. Attempts by our laboratory to block PD-1 or B7-H1 concurrent with administration of immunotherapy did not result in the desired expansion of CD4+ T cells (unpublished observation). This was attributed to the degree of PD-1 and B7-H1 upregulation which would have required very large quantities of antibody to inhibit signaling. Studies in PD-1 or B7-H1 knockout animals may be the only way to determine the degree of importance this pathway plays on the efficacy of immunotherapy.

While PD-1 has only recently been of interest with respect to its strong inhibitory properties in the area of tumor immunotherapy, previous interest has been on other surface receptors with known inhibitory and or proapoptotic effects on T cells. These include members of the TNF family of receptors and ligands. Members of this family have important roles in the tight immunological regulation of activated T cells, including activation-induced cell death (AICD). AICD is a process through which activated T cells are tightly regulated and can happen through the interaction of Fas/FasL or DR5/TRAIL and may require the presence of IFNγ.[13,38,39] Due to its promotion of AICD, TRAIL has been implicated in the control of immunological memory. It has been shown to mediate removal of antigen-specific CD8+ T cells that have not been properly primed by CD4+ T cells.[40] The majority of data have focused on the role of FasL and TRAIL on CD8+ T cell AICD.[41,42] Our findings summarized here regarding both the direct effects of IFNγ on CD4+ T cells[7] and the concurrent upregulation of PD-1/B7-H1 after immunotherapy (submitted) complement these previous findings by illustrating further subset-specific complexity in the regulation of the T cell compartment.

In this review, we have discussed the importance of PD-1 and B7-H1 in response to immunotherapy. PD-1 and B7-H1 are likely to be key immune markers involved in the imbalance of normal lymphocyte ratios that is observed after the administration of immunotherapy. This imbalance has a twofold detrimental effect on lasting antitumor responses. First, a significant reduction in the CD4:CD8 balance does not allow for proper CD8+ T cell priming. Therefore, immunological memory is not retained to tumor antigens. Second, PD-1 is selectively upregulated on the surface of CD4+ Tconv cells and not on Treg cells. This allows for Treg cells to expand following immunotherapy whereas Tconv cells cannot. This may have a detrimental effect not only on the ability of the primary responding cells, but also a compounding inhibitory effect on the generation of functional T cell memory. Furthermore, this effect is dependent on IFNγ and therefore has applications in any immunotherapeutic regimen that utilizes IFNγ to elicit an immunological response. Removal of Treg cells with or without the use of blocking antibodies for PD-1 and B7-H1 may be the best way to maximize the application of IFNγ-dependent immunotherapy. Further studies relying on the application of immunotherapy should consider this strong immunological opposition with respect to limiting the efficacy of treatment in cancer.

Acknowledgments We thank Dr. Ruth Gault for her help with figures and Dr. Lisbeth Welniak for her help with reviewing the manuscript. This work was supported in part by R01 CA 95572.

References

1. June CH. Adoptive T cell therapy for cancer in the clinic. *J Clin Invest* 2007;**117**:1466–1476
2. June CH. Principles of adoptive T cell cancer therapy. *J Clin Invest* 2007;**117**:1204–1212
3. Robbins PF, Dudley ME, Wunderlich J, . Cutting edge: persistence of transferred lymphocyte clonotypes correlates with cancer regression in patients receiving cell transfer therapy. *J Immunol* 2004;**173**:7125–7130

4. Morgan RA, Dudley ME, Wunderlich JR, . Cancer regression in patients after transfer of genetically engineered lymphocytes. *Science* 2006;**314**:126–129
5. Kershaw MH, Westwood JA, Parker LL, . A phase I study on adoptive immunotherapy using gene-modified T cells for ovarian cancer. *Clin Cancer Res* 2006;**12**:6106–6115
6. Dudley ME, Wunderlich JR, Robbins PF, . Cancer regression and autoimmunity in patients after clonal repopulation with antitumor lymphocytes. *Science* 2002;**298**:850–854
7. Berner V, Liu H, Zhou Q, . IFN-gamma mediates CD4+ T-cell loss and impairs secondary antitumor responses after successful initial immunotherapy. *Nat Med* 2007;**13**:354–360
8. Murphy WJ, Welniak L, Back T, . Synergistic anti-tumor responses after administration of agonistic antibodies to CD40 and IL-2: coordination of dendritic and CD8+ cell responses. *J Immunol* 2003;**170**:2727–2733
9. Barker BR, Parvani JG, Meyer D, Hey AS, Skak K, Letvin NL. IL-21 induces apoptosis of antigen-specific CD8+ T lymphocytes. *J Immunol* 2007;**179**:3596–3603
10. Damle NK, Leytze G, Klussman K, Ledbetter JA. Activation with superantigens induces programmed death in antigen-primed CD4+ class II+ major histocompatibility complex T lymphocytes via a CD11a/CD18-dependent mechanism. *Eur J Immunol* 1993;**23**:1513–1522
11. Dai Z, Arakelov A, Wagener M, Konieczny BT, Lakkis FG. The role of the common cytokine receptor gamma-chain in regulating IL-2-dependent, activation-induced CD8+ T cell death. *J Immunol* 1999;**163**:3131–3137
12. Zhang J, Bardos T, Shao Q, . IL-4 potentiates activated T cell apoptosis via an IL-2-dependent mechanism. *J Immunol* 2003;**170**:3495–3503
13. Zhang J, Xu X, Liu Y. Activation-induced cell death in T cells and autoimmunity. *Cell Mol Immunol* 2004;**1**:186–192
14. Bartholdy C, Kauffmann SO, Christensen JP, Thomsen AR. Agonistic anti-CD40 antibody profoundly suppresses the immune response to infection with lymphocytic choriomeningitis virus. *J Immunol* 2007;**178**:1662–1670
15. Tong AW, Stone MJ. Prospects for CD40-directed experimental therapy of human cancer. *Cancer Gene Ther* 2003;**10**:1–13
16. Ishida Y, Agata Y, Shibahara K, Honjo T. Induced expression of PD-1, a novel member of the immunoglobulin gene superfamily, upon programmed cell death. *EMBO J* 1992;**11**:3887–3895
17. Martin-Orozco N, Wang YH, Yagita H, Dong C. Cutting edge: programmed death (PD) ligand-1/PD-1 interaction is required for CD8+ T cell tolerance to tissue antigens. *J Immunol* 2006;**177**:8291–8295
18. Goldberg MV, Maris CH, Hipkiss EL, . Role of PD-1 and its ligand, B7-H1, in early fate decisions of CD8 T cells. *Blood* 2007;**110**:186–192
19. Zhang JY, Zhang Z, Wang X, . PD-1 up-regulation is correlated with HIV-specific memory CD8+ T-cell exhaustion in typical progressors but not in long-term nonprogressors. *Blood* 2007;**109**:4671–4678
20. Chemnitz JM, Eggle D, Driesen J, et al. RNA-fingerprints provide direct evidence for the inhibitory role of TGF{beta} and PD-1 on CD4+ T cells in Hodgkin's lymphoma. Blood 2007
21. Colley DG, Sasser LE, Reed AM. PD-L2+ dendritic cells and PD-1+ CD4+ T cells in schistosomiasis correlate with morbidity. *Parasite Immunol* 2005;**27**:45–53
22. Hatachi S, Iwai Y, Kawano S, . CD4+ PD-1+ T cells accumulate as unique anergic cells in rheumatoid arthritis synovial fluid. *J Rheumatol* 2003;**30**:1410–1419
23. Flies DB, Chen L. The new B7s: playing a pivotal role in tumor immunity. *J Immunother* 2007;**30**:251–260
24. Okazaki T, Iwai Y, Honjo T. New regulatory co-receptors: inducible co-stimulator and PD-1. *Curr Opin Immunol* 2002;**14**:779–782
25. Dong H, Strome SE, Salomao DR, . Tumor-associated B7-H1 promotes T-cell apoptosis: a potential mechanism of immune evasion. *Nat Med* 2002;**8**:793–800
26. Keir ME, Francisco LM, Sharpe AH. PD-1 and its ligands in T-cell immunity. *Curr Opin Immunol* 2007;**19**:309–314
27. Grakoui A, John Wherry E, Hanson HL, Walker C, Ahmed R. Turning on the off switch: regulation of anti-viral T cell responses in the liver by the PD-1/PD-L1 pathway. *J Hepatol* 2006;**45**:468–472

28. Khoury SJ, Sayegh MH. The roles of the new negative T cell costimulatory pathways in regulating autoimmunity. *Immunity* 2004;**20**:529–538

29. Blazar BR, Carreno BM, Panoskaltsis-Mortari A, . Blockade of programmed death-1 engagement accelerates graft-versus-host disease lethality by an IFN-gamma-dependent mechanism. *J Immunol* 2003;**171**:1272–1277

30. Hori J, Wang M, Miyashita M, . B7-H1-induced apoptosis as a mechanism of immune privilege of corneal allografts. *J Immunol* 2006;**177**:5928–5935

31. Picca CC, Larkin J, III, Boesteanu A, Lerman MA, Rankin AL, Caton AJ. Role of TCR specificity in CD4+ CD25+ regulatory T-cell selection. *Immunol Rev* 2006;**212**:74–85

32. Barao I, Hanash AM, Hallett W, . Suppression of natural killer cell-mediated bone marrow cell rejection by CD4+ CD25+ regulatory T cells. *Proc Natl Acad Sci USA* 2006;**103**: 5460–5465

33. Beswick EJ, Pinchuk IV, Das S, Powell DW, Reyes VE. B7-H1 expression on gastric epithelial cells after *Helicobacter pylori* exposure promotes the development of CD4+ CD25+ FoxP3+ regulatory T cells. Infect Immun 2007

34. Raimondi G, Shufesky WJ, Tokita D, Morelli AE, Thomson AW. Regulated compartmentalization of programmed cell death-1 discriminates CD4+ CD25+ resting regulatory T cells from activated T cells. *J Immunol* 2006;**176**:2808–2816

35. Tarhini AA, Agarwala SS. Interleukin-2 for the treatment of melanoma. *Curr Opin Investig Drugs* 2005;**6**:1234–1239

36. Maker AV, Phan GQ, Attia P, . Tumor regression and autoimmunity in patients treated with cytotoxic T lymphocyte-associated antigen 4 blockade and interleukin 2: a phase I/II study. *Ann Surg Oncol* 2005;**12**:1005–1016

37. Tsushima F, Yao S, Shin T, . Interaction between B7-H1 and PD-1 determines initiation and reversal of T-cell anergy. *Blood* 2007;**110**:180–185

38. Green DR, Droin N, Pinkoski M. Activation-induced cell death in T cells. *Immunol Rev* 2003;**193**:70–81

39. Refaeli Y, Van Parijs L, Alexander SI, Abbas AK. Interferon gamma is required for activation-induced death of T lymphocytes. *J Exp Med* 2002;**196**:999–1005

40. Janssen EM, Droin NM, Lemmens EE, . CD4+ T-cell help controls CD8+ T-cell memory via TRAIL-mediated activation-induced cell death. *Nature* 2005;**434**:88–93

41. Aggarwal BB. Signalling pathways of the TNF superfamily: a double-edged sword. *Nat Rev Immunol* 2003;**3**:745–756

42. Smyth MJ, Takeda K, Hayakawa Y, Peschon JJ, van den Brink MR, Yagita H. Nature's TRAIL – on a path to cancer immunotherapy. *Immunity* 2003;**18**:1–6

New Insights into Classical Costimulation of CD8+ T Cell Responses

Christine M. Bucks and Peter D. Katsikis(✉)

1 Introduction

Nearly 40 years ago a theory was put forth postulating that an immune cell must receive two signals during activation to discriminate self- from nonself-antigen. One signal must be through an antigen-specific receptor and the second signal must be through a costimulatory receptor which would indicate that the antigen was indeed foreign and that a response was required. This theory came to be known as the two-signal theory and was long accepted as a key paradigm of immunology. Since its inception, it has sparked a wildfire of research in pursuit of the identification and characterization of such second signal molecules. Today, more than ten costimulatory signals have been identified, both stimulatory and inhibitory, and a multiple signal model has replaced the oversimplified two-signal theory. Here, we will provide an introduction to the members of the classical CD28 family, with a primary focus on CD28. We will review the function of classical CD28 costimulation in the generation of primary and memory antiviral CD8+ T cell responses. Particular emphasis will be placed on understanding the contribution of CD28 costimulation to efficient antiviral CD8+ T cell memory responses. Until recently, the importance of CD28 signaling in such memory CD8+ T cell responses has remained unappreciated. Finally, we will discuss the implications of these new insights into the role of CD28 as they relate to the future of vaccine design, tumor suppression, and autoimmunity.

2 CD28 and TNF/TNFR Costimulatory Molecules

Efficient generation of adaptive immune responses requires the presence of an antigen-specific signal and one or more costimulatory signals. When this concept was first brought to light by Bretscher and Cohn in 1970,[1] the theory stated that B

Peter D. Katsikis
Department of Microbiology and Immunology, Drexel University College of Medicine
2900 Queen Lane, Philadelphia, PA 19129, USA

S.P. Schoenberger et al.(eds.) *Crossroads between Innate and Adaptive Immunity II*,
doi: 10.1007/978-0-387-79311-5_9, © Springer Science+Business Media, LLC 2009

cells must distinguish "paralyzing" antigen from foreign antigen through the presence of a second activating signal. This theory was later expanded to include T cells.[2] During the pursuit to identify the second activating signal, a monoclonal antibody against a 45-kDa glycoprotein was generated; the protein target of this monoclonal antibody would later be cloned and come to be known as CD28.[3] Early studies using PHA stimulated T cells revealed that mitogen-induced proliferation could be enhanced by the addition of this monoclonal antibody to cell culture.[4] Later, it was shown that the effects of anti-CD28 were not limited just to mitogen stimulation, but could also be observed during CD3-mediated T cell stimulation.[5-8] Furthermore, the increased proliferation was shown to be directly mediated through CD28, in a pathway which was distinct from CD3 signaling cascade.[9,10] In the years since, a multitude of costimulatory molecules belonging to both the classical CD28 and alternative TNF/TNFR superfamilies have been identified. Our understanding of costimulatory molecules has evolved and now even includes accessory molecules which have an inhibitory rather than stimulatory function.

The identification of CD28 led to the discovery of a series of other costimulatory molecules which have been grouped together on the basis of shared structural homology. Members of the CD28 superfamily all contain a variable Ig-like extracellular domain and a short cytoplasmic tail. To date, there are five confirmed members of the classical CD28 family. The receptor: ligand pairs are as follows: CD28:B7-1 or B7-2, CTLA4:B7-1or B7-2, ICOS:B7h, PD-1:B7H-1 or B7-DC, and BTLA:HVEM. Two additional molecules which are believed to be CD28 superfamily ligands, but are as yet unpaired with receptors, are B7H3 (also known as B7x) and B7H4 (also known as B7S1). A second group of alternative costimulatory molecules, the TNF/TNFR superfamily, can be distinguished from CD28 members by the presence of a more complex cytoplasmic tail. Members of the TNF/TNFR family can be subdivided into three groups: those containing cytoplasmic death domains, those lacking a death domain but containing decoy receptors, and those which lack a death domain but contain a TRAF motif. TNF/TNFR receptor:ligand pairs which have known costimulatory functions include OX-40:OX40L, CD27: CD70, 4-1BB:4-1BBL, CD30:CD30L, GITR:GITRL, and HVEM:Light.

Further discussion will be restricted only to those members of the classical costimulatory CD28 family, with special emphasis on CD28. However, our discussion would be incomplete without providing at least a cursory overview of the expression pattern and function of the other members of the CD28 family. Understanding these basics about the CD28 family members is important as the function of CD28 can be largely impacted by the presence or absence of its other family members. Both CTLA4 and ICOS are structural homologs of CD28, yet they exhibit unique, nonredundant functions upon stimulation. CTLA4 competes for binding of the same ligands as CD28, and as such is not expressed on resting or newly activated T cells. Rather, expression is largely restricted to fully activated T cells and regulatory cells.[11] CTLA4 is estimated to bind B7-1 and B7-2 with a 10–20-fold greater affinity than CD28.[12] Binding by B7-1 or B7-2 leads to the disruption of lipid rafts,[13] the interruption TCR signaling, and thus inhibition of T cell proliferation.[14,15] In contrast, the inducible costimulator, ICOS, does not share ligands with CD28 but rather binds its own unique ligand, B7-h. Signaling through ICOS augments proliferation, antibody response, and cytokine production.[16–19] With regard to viral infection, in vivo studies indicate a particularly important role for ICOS signaling during development of primary antibody responses and maintenance of late stage primary CD8+ T cell responses during LCMV, VSV, and influenza.[20]

The more recently identified members of the CD28 family, Programmed Death-1 (PD-1) and B and T cell lymphocyte attenuator (BTLA), both exhibit largely inhibitory activity. PD-1 has at least two known ligands PD-L1 (B7H-1) and B7-DC, and is expressed on both T and B cells. Signaling through PD-1 has been shown to be involved in peripheral tolerance[21] and in the regulation of antiviral CD8+ T cell responses during chronic infection.[22–26] Significantly less is known about the newest member of the CD28 family, BTLA. BTLA is constitutively expressed at low levels on T cells and can be inducibly upregulated on activated B and T cells. It has been suggested to play an inhibitory role in the development of adaptive immune responses.[27] Recent in vivo evidence indicates that BTLA signaling can inhibit CTL maturation and memory generation.[28] It is interesting to consider the extensive effect on the outcome of T cell function, which occurs merely by changing the type of costimulatory molecule which is signaled. The uniqueness of each costimulatory molecule emphasizes the fine degree of control that an activated cell must be maintained under and reminds us of the precarious balance between stimulating adequate immune responses and breaking tolerance.

3   CD28 Expression and Intracellular Signaling

CD28 is observed from the very earliest stages of T cell thymocyte development and continues to be found, albeit at different levels, throughout the lifespan of the T cell. In fact, the presence of CD28 on early thymocytes may represent one of the earliest indicators of the importance of this costimulatory molecule in both the survival and function of T cells. As a developing progenitor cell enters the thymus it becomes committed to the T cell lineage. It will undergo successive rounds of both positive and negative selection to yield a functional, nonautoreactive T cell which will be released to the periphery. In the thymus, CD28 is expressed on about 50% of developing thymocytes in humans[29] and on 90% of thymocytes in mice.[30] Its ligand, B7, is expressed on thymic epithelia and dendritic cells (DCs) of the corticomedullary and medullary thymus.[31,32] A developing thymocyte transits through three basic stages: double negative T cell (DN) to double positive (DP) and finally to single positive (SP). Once a thymocyte has transitioned from DN to DP it begins the process of becoming an immature SP T cell. Numerous studies indicate that this transition from DP to SP requires CD28 signaling[33–35] and that survival or death of the DP cell is dependent on the intensity of that signal.[33] Based on these findings CD28 appears to play a significant role in thymic selection and thus can be considered important in central tolerance. In light of these data, it is surprising that CD28-deficient mice contain a peripheral repertoire of CD4+ and CD8+ T cells which appear to have no obvious defects.[36] In fact, the only defect which has been identified is the loss of the CD4+ regulatory T cells.[37,38] These two findings at first seem to conflict with each other. However, closer examination shows that absence of CD28 does not prevent T cell development but, rather, negatively affects the ability of these cells to differentiate and survive. Walunas et al.[39] elegantly illustrated this, using mixed chimera irradiated mice which were reconstituted with immature single positive T cells derived from CD28-competent and CD28-deficient mice. When given in different ratios, the T cells from the CD28-competent mice consistently proliferated more and survived better than their CD28-deficient counterparts. Only when CD28-deficient immature T cells were given in excess were they able to overcome the CD28-sufficient T cells.[39] Based on these findings, one can conclude that CD28 plays an important qualitative role in the development of functional peripheral T cells.

Having successfully completed multiple rounds of thymic selection, the T cell is released into the periphery. Analysis of adult human peripheral blood T cells shows that CD28 is constitutively expressed on almost all CD4+ T cells and about 50% of CD8+ T cells.[40] Additionally CD28 is also reported to be expressed on some human plasmablasts and plasma cells.[41,42] CD28 expression is maintained throughout early life, but as an individual ages its expression decreases.[43–45] A similar loss of CD28 expression is also observed in individuals afflicted with chronic infection or autoimmune disease.[46] In the periphery, the ligands for CD28, B7-1, and B7-2 (CD80 and CD86) are widely expressed on antigen-presenting cells such as dendritic cells and macrophages.[47,48] Similar to CD28, CD86 is constitutively expressed at low levels and increases following activation.[47,48] In contrast, CD80 is only inducibly expressed following activation of the APC.[48]

CD28 is expressed as a homodimer and when unligated has no intrinsic activity. However when the TCR is properly engaged and CD28 binds B7-1 or B7-2, the activities attributed to CD28 signaling are numerous. These include, but are not limited to, an increase in glucose metabolism,[49] increased cytokine and chemokine production,[50] increased IL-2Rα expression,[51] increased IL-2 production,[9,51,52] increased rate of entry into the cell cycle,[53,54] and enhanced resistance to apoptosis.[55,56] Signaling through CD28 also appears to lower the threshold of activation by decreasing the number of TCRs which are required to be engaged to activate a cell.[57,58] When the TCR is engaged in the absence of CD28 costimulation, the responding T cell becomes tolerized and is functionally incapable of responding to further antigen stimulation.[59–61]

In order for these events to occur, signals transduced through the CD28 receptor must activate downstream kinases, adaptor molecules, and transcription factors (Fig. 1). Activation of target molecules occurs either synergistically with TCR-mediated events or through a unique pathway which is independent of the CD3/TCR signaling complex.[9,10] One of the first events to occur following CD28 ligation is recruitment of SRC and TEC family kinases to four highly conserved intracellular tyrosine residues on the 41aa cytoplasmic tail. These tyrosine residues are phosphorylated, and subsequently bind phosphatidylinositol 3-kinase (PI3K) or growth-factor-receptor-bound-protein-2 (GRB2) family proteins. Downstream of PI3K and

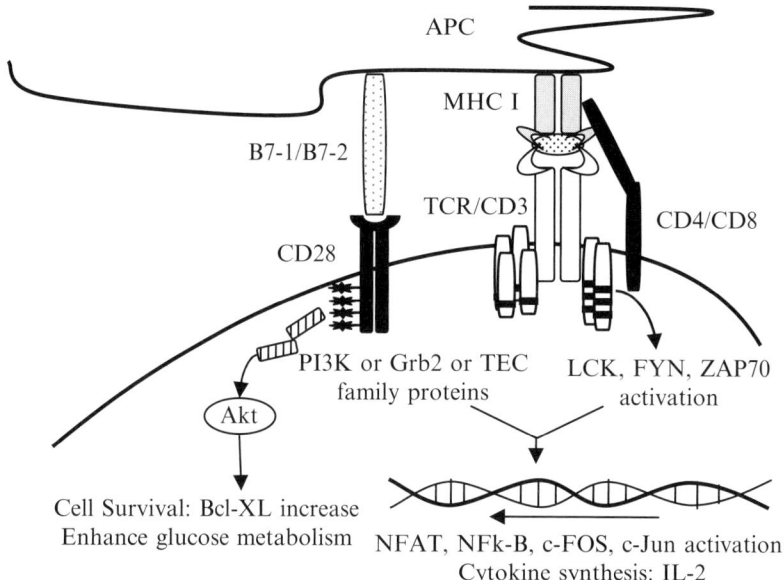

Fig. 1 CD28 signaling leads to multifaceted downstream cellular effects. Signaling through CD28 leads to the recruitment of PI3K, Grb2, and TEC family proteins. Activated molecules either synergize with TCR signaling pathway and amplify/stabilize cytokine transcription, or act through an independent Akt pathway to increase glucose metabolism and cell survival factors

GRB2, several pathways may be activated, including Akt, NFk-B, and NFAT. One of the classically associated downstream effects of CD28 costimulation is the upregulation of the IL-2Rα chain and increased transcription of IL-2.[9,51,52] The production of IL-2 by the activated T cell leads to both autocrine and paracrine effects, promoting proliferation of both self and bystander cells. Production of IL-2 represents one of the synergistic products generated by both TCR/CD3 and CD28 signaling. Signaling through the TCR-independent pathway leads to tyrosine residue phosphorylation, PI3K recruitment, and Akt activation. Activation of Akt leads to an increase in the Glut1 glucose transporter and subsequent glycogen synthesis.[49] The increase in glucose metabolism reflects the cell's growing need for nutrients as it becomes more energetically active and enters the cell cycle. Entry into cell cycle is partially facilitated by another CD28-dependent event. Cyclin-dependent kinase inhibitor, $P27^{kip1}$, is responsible for maintaining a cell in the quiescent state of the cell cycle, stage G_0.[62] Following CD28-mediated activation of the PI3K pathway, $P27^{kip1}$ is downregulated and both D and E cyclins of the G1 phase become activated. The CD28 stimulated cell now enters the cell cycle in an IL-2-independent manner.[53,54] In concert with promoting glucose metabolism and proliferation, CD28 signaling functions also alters the cell's susceptibility to apoptosis. As a consequence of CD28 signaling, NFk-B is activated and both transcription and translation of the antiapoptotic molecules Bcl-X_L and c-FLIPshort are increased.[55,56,63,64] The enhanced expression of these molecules provides yet another layer of insurance that the signaled T cell will be activated and protected from activation-induced cell death during response to foreign antigen.

Much of the current knowledge of the CD28 signaling pathway is based on studies using naïve T cells. As we shall see below, CD28 also provides important signals to memory T cells. There are as yet no reported studies examining the signaling events which occur during the reactivation of memory T cells. It remains possible that the signaling pathway activated in memory T cells could be distinctly different from that which we observe in naïve cells. Understanding the differences in signaling between naïve and memory T cells could have important implications for CD28-based therapeutics and represents an unexplored frontier in the field. All the work discussed above focused on determining the function and signaling properties of CD28 in antigen-mediated T cell stimulation in vitro and in vivo. Our discussion moves now to analysis of CD28 function in the generation of primary and memory responses to viral infection. Here, in addition to antigen and CD28 availability, the relative antigen load, viral tropism, and the inflammatory state that is associated with the viral infection must also be considered. As we shall see, all these components can change the requirement for CD28 in the development of antiviral responses and thus the interpretation of CD28's functional significance.

4   CD28 and Primary Antiviral Immune Responses

Understanding the function of CD28 on antiviral CD8+ T cells during the generation of a primary immune response requires the examination of findings from a variety of different infection models, including lymphocytic choriomeningitis virus (LCMV), vesicular stomatitis virus (VSV), murine gamma herpes virus (MHV-68), and influenza virus. Studying a variety of viral infections is important because, as we will see, the intrinsic properties of a particular viral infection can significantly influence the need for, and thus role of, CD28 costimulation in the generation of primary T cell responses.

The importance of CD8+ T cells in the resolution of viral infection is widely accepted. During viral infection circulating dendritic cells present viral antigen to CD8+ T cells in the local draining lymph nodes. The complex nature of these interactions between the antigen-presenting DC and the naïve CD8+ T cell has been extensively studied and is reviewed elsewhere.[65,66] Importantly it is during this initial interaction with the CD8+ T cell that the decision is made whether or not to expand or become anergized. One of the crucial determinants in this decision is the presence or absence of CD28 costimulation. If both signal one, TCR stimulation, and signal two, CD28 costimulation, are achieved the cell undergoes multiple rounds of division and transits to the site of infection. Here the CD8+ T cell exerts its cytolytic- and cytokine-producing functions, clearing the host of infection. Concurrent with resolution of the viral burden, the effector CD8+ T cell population undergoes a steady contraction until a small stable memory pool is formed. This CD8+ T cell memory population remains poised and ready to respond in the event of secondary insult to the host. This generalized model describes the events occurring in many viral infections, but fails to highlight the importance of CD28 costimulation under different conditions. We will now examine specific viral infections and explore the varying requirement for CD28 costimulation in the generation of anti-viral T cell responses.

LCMV infection provides an excellent example of how the requirement for CD28 costimulation can be altered due to intrinsic properties of the virus. One of the first tools used to understand the requirement of CD28 signaling for the development of primary CD8+ T cell responses was LCMV infection of CD28 knockout mice. LCMV is a natural mouse pathogen which can be an acute or chronic infection depending on the dose and strain of virus used and the immune competency state of the host. The ability to study the effect of altering the severity and chronicity of infection makes it an attractive choice of viral model system to study. Studies with LCMV initially showed that an efficient primary CD8+ T cell response could be generated in the absence of CD28 costimulation. CD28 knockout mice (CD28−/−) were infected with LCMV, and despite the absence of CD28 signaling, virus CD8+ T cells expanded and viral burden was eliminated at levels comparable to wild-type controls. Additionally comparable CD8+ T cell expansion was observed against all measured epitopes of LCMV, including subdominant epitopes.[67,68] These studies led to the early suggestion that CD28 signaling was dispensable in the generation of primary antiviral CD8+ T cell responses. In fact the reality of the CD28 requirement turned out to be much more complex than these studies suggest.

In striking contrast to LCMV infection, subsequent studies examining primary infection of either CTLA4-Ig-treated, CD80/86 or CD28 knockout mice with other viruses such as VSV,[69] MHV-68,[70] and influenza[71–74] indicated that CD28 was required for primary expansion of antiviral CD8+ T cells. In one of the earliest CTLA4-Ig blocking studies, using influenza, Lumsden et al.[71] identified that the loss of CD28 signaling negatively impacted both CD4+ and CD8+ T cells. In this study, there was a significant decrease in the production of antiviral antibodies, decreased expansion of virus-specific CTLs, and a loss of IFN-γ and cytotoxic function by those cells which did expand.[71] Ultimately these CTLA4-Ig-blocked mice resolved the infection, but it was delayed in comparison to controls.[71] Halstead et al.[73] also showed that both dominant and subdominant CD8+ T cell responses against influenza virus are greatly reduced in CD28 knockout mice. In a comple-mentary study by Bertram et al.,[72] influenza-infected CD28−/− mice exhibited sub-stantially decreased expansion of virus-specific CD8+ T cells at the peak of the primary response whether virus was delivered intraperitoneally or intranasally. Paradoxically, the study with LCMV seems to clearly indicate that CD28 is dispen-sable for generation of primary antiviral CD8+ T cell responses, yet a number of in vivo studies using other viral infections indicate that CD28 is critical for primary CD8+ T cell expansion and function. So why do these disparities exist?

The answer to these questions most likely lies in the details of the viral infections themselves and provides important information about the conditions under which CD28 costimulation is required. In vitro systems using anti-CD3 stimulation of T cells have shown that a response, albeit a weak one, can be generated in the absence of costimulation. The addition of anti-CD28 mAb in these cultures enhances the proliferation and function of these cells.[5–8] The physiological relevance of in vitro studies is always questionable; however, it is clear that in vitro with high anti-CD3 concentrations once can overcome the CD28 costimulation requirement. At lower anti-CD3 concentrations, however, the addition of CD28 costimulation has profound effects on proliferation. These studies were the foundation upon which the idea that CD28 costimulation is required for primary CD8+ T cell generation was built. More sophisticated analysis of TCR stimulation events in these in vitro cultures led to the strength of signal theory. This theory directly addresses the reason for the disparity in findings between the LCMV and VSV, MHV, and influenza studies. In back-to-back publications, two independent groups showed that if sufficiently high levels of TCR stimulation were obtained, the need for costimulation could be overcome.[57,75] Viola et al.[57] elegantly showed in in vitro studies that independent of the nature of the TCR stimuli, TCR stimulation must exceed a minimum threshold to achieve complete activation of a T cell clone. However, in the presence of CD28 costimu-lation, that threshold is significantly lowered. Kundig et al.[75] utilized LCMV infec-tion and showed that the disparity in requirement for CD28 in primary LCMV infection vs. VSV infection was due to differences in TCR signal duration. Indeed, LCMV is the only one of all the viruses examined whose natural host is the rodent. Therefore it replicates much more rapidly and extensively than any of the other viruses examined. As a result, antigen presentation persists for a longer period of time and at higher levels, providing a strong and sustained TCR signal which over-comes the need for CD28 costimulation (Fig. 2).

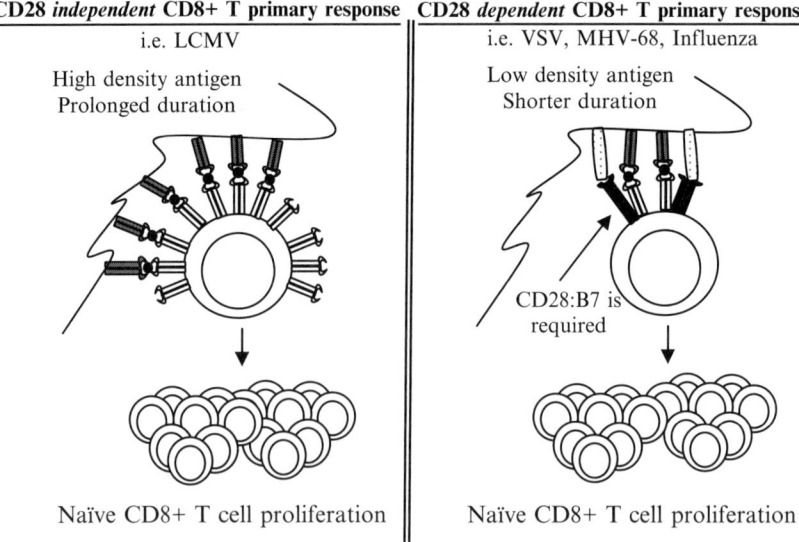

Fig. 2 Requirement for CD28 in primary antiviral CD8+ T cell responses is dependent on strength of TCR signal. Requirement for CD28 costimulation varies by the infection model studied. LCMV infection appears to generate CD8+ T cell primary responses independent of CD28, while VSV, MHV-68, and influenza all require CD28 for optimal antiviral CD8+ T cell response. This discrepancy in requirement is attributed to variations in viral load, tissue tropism, and inflammation seen in the previously mentioned infections

In the end, one can conclude that CD28 enhances the generation of antiviral CD8+ T cell primary responses. Certainly in many cases, it is an absolute requirement for the generation of efficient primary CD8+ T cell responses. It is only when the antigen load reaches extremely high levels or persists for an extended period of time that this requirement can be overcome and a one signal through the TCR is sufficient for T cell proliferation and function. Given that CD28 signaling has been shown to be important for the development of primary antiviral CD8+ T cell response it is reasonable to question if CD28 is important for the generation of memory CD8+ T cells, or if memory is hardwired to act even in the absence of classical CD28 costimulation.

5   CD28 and Memory Antiviral Immune Responses

Memory CD8+ T cells represent a critical population of T cells which are capable of responding to viral infection. Unlike primary effector cells, memory cells exist as an antigen educated pool of cells which rapidly proliferate and produce cytokine upon detecting antigen. One of the major contributing factors to the rapidity of memory T cell responses is their potentially affinity for antigen and thus lower threshold of activation.[76] Given that the strength of TCR signaling and predetermined threshold of activation can affect the need for costimulation, it is reasonable to question whether or not memory CD8+ T cells have a requirement for CD28 costimulation during reactivation. Interestingly, CD28 signaling requirements in memory CD8+ T cell responses have been much less well studied than those for primary response generation. In fact the widely held belief that memory responses are CD28 costimulation independent is based on very few studies which predominantly utilized in vitro systems of restimulation[77,78] or CD28 knockout mice.[68,79,80] These studies used a variety of different priming and rechallenge systems but the majority shared the common finding that memory T cell responses, both CD4+ and CD8+, were independent of CD28 costimulation.[78,79,81,82]

One of the first studies to make the transition from in vitro to in vivo study of CD28 costimulation requirements by memory CD8+ T cells was performed using murine influenza A virus.[74] In this study, the generation of both primary and secondary CD8+ T cell responses to influenza was studied in CD28 knockout or B7 monoclonal antibody (mAb)-treated mice. Consistent with others,[71,72] development of the primary antiviral CD8+ T cell response was inhibited in the absence of CD28 or its B7 ligand.[74] However the unique feature of this study was that the impact of CD28 signaling on the secondary CD8+ T cell response was also explored. Here, memory is quantified on the basis of in vitro recall responses, specifically, ex vivo cytotoxicity ^{51}Cr-release assay to influenza antigen. An important caveat of this study is that a true secondary in vivo infection was never used, and thus other factors, such as antigen load and cytokine environment, which influence the need for costimulation are not considered. To study the reactivation of a pre-existing memory CD8+ T cell pool, memory was generated in either CD28-sufficient or

CD28 knockout mice. The mice were rested for more than three months, and then the function of the resultant CD8+ memory T cells was assessed by ^{51}Cr-release assay after a 5-day in vitro expansion. Independent of the initial priming conditions, CD28 was shown to be required for maximal killing by these memory CD8+ T cells. In absence of CD28 signaling, cytotoxic responses could be observed, but the response was dramatically reduced in comparison to controls. In short, this is one of the first studies to indicate that CD28 signaling was important not only for efficient activation of primary CD8+ T cells, but that it also was critical for the in vitro reactivation of memory CD8+ T cells.

Soon after this study the advent of tetramer technology occurred[83] along with the widespread use of CD8+ TCR transgenic mice. These advancements induced an explosion of in vivo research related to the study of CD8+ memory T cells. Prior to the development of such tools, study of virus-specific CD8+ T cells was limited to functional assays such as ^{51}Cr-release and antigen-specific proliferation assays, as discussed above. While useful, these assays were limited in their ability to track small populations of circulating CD8+ memory T cells, and therefore a significant obstacle existed when trying to understand the role of CD28 costimulation in memory T cell reactivation. Some of the early knowledge about CD28 signaling on memory CD8+ T cells, in vivo, was gained peripherally in studies which used adoptive transfer of transgenic CD8+ T cells in the pursuit of a broader scope of knowledge about memory CD8+ T cells themselves. A prime example of this is the study of CD8+ T cell memory generated in VSV-OVA-infected mice.[79] Here, OVA-specific OT-I TCR transgenic CD8+ T cells were transferred to naïve CD28 intact mice. Recipient mice were then infected with VSV-OVA virus and orally rechallenged by OVA feeding in the presence or absence of CTLA4-Ig. Examination of memory CD8+ T cells after oral rechallenge indicated that memory CD8+ T cells functioned independent of CD28 costimulation.[79] This conclusion was reached, based on the findings that both the degree of blastogenesis (size increase) and ex vivo cytolytic function were comparable in CTLA4-Ig-treated and control mice. These findings are in agreement with previously published in vitro studies,[78,81,82,84] but are in contrast to the in vivo influenza study discussed above.[74] Although this study concluded that CD8+ T cell memory responses occurred independent of CD28 signaling, this was not proven in a comprehensive way. The conclusions were based on basic blastogenesis analysis of the antigen-specific population. As we shall see below, the size of memory CD8+ T cells is not affected when cells fail to receive CD28 costimulation; however, there is an accompanying cell cycle arrest.[85] The ex vivo cytolytic assay on which the conclusion was also based, had the number of OT-I cells adjusted. Therefore it did not demonstrate that there were fewer expanded secondary OT-I cells, only that on a per-cell basis, killing is not affected by blocking CD28 costimulation. Additionally, this study and the others discussed, failed to examine the requirement for CD28 signaling in an in vivo secondary infection. Moreover analysis of the kinetics of the secondary response, quantitation of the response, and cytokine production capacity of these memory CD8+ T cells is absent. Without this more in-depth analysis of the function of these cells, it is premature to conclude that memory is independent of CD28 costimulation.

As we have seen in the primary response, the requirement for CD28 signaling in the activation of CD8+ T cells can be largely impacted by the type of viral system studied. Similar to the primary CD8+ T cell response, LCMV-specific memory CD8+ T cell responses in CD28 knockout mice also seem to be reactivated independent of CD28 costimulation.[68] A recent report by Suresh et al. showed that consistent with previous studies,[67] in LCMV-infected CD28−/− mice, the primary CD8+ T cell responses exhibited a strong activation profile and primary expansion. Although the CD28−/− population was reduced in size in comparison to controls, it still represented a sizeable population and exhibited substantial CTL function.[67] The LCMV-specific CD8+ T cell memory population was examined out to be greater than 250 days postprimary infection and observed to be only slightly decreased in size relative to wild-type controls. When wild-type or CD28−/− memory mice were rechallenged with a lethal dose of LCMV, all mice survived infection while all naïve controls in the study died. One interpretation of these data is that, during LCMV infection, memory CD8+ T cells are capable of functioning independent of CD28 costimulation.

In light of the previously discussed studies it appears that there is overwhelming evidence indicating that CD28 costimulation is dispensable for CD8+ T cell memory responses. And in fact this is a generally accepted paradigm in immunology. However we should stress that all these previously discussed studies have examined CD28 costimulation requirements under conditions where T cell stimulus has been extensively manipulated. That is, in the in vitro studies, the requirement for costimulation may have been overcome due to the strength of TCR signaling. In the previously described studies peptide was exogenously loaded onto cultured antigen-presenting cells, effectively over riding normal antigen processing and presentation MHC class I pathway. Finally we must consider LCMV infection data, which arguably utilized the most physiologically relevant antigen presentation system and functional readout of memory. While these studies do indicate that CD8+ T cell memory reactivates independent of CD28 costimulation, we stress several caveats which may alter the interpretation of the data. First, the CD8+ T cell memory is generated in CD28−/− mice. As we have seen CD28 and its ligands are highly expressed in the thymus.[29–32] While CD28 knockout mice appear to have normal peripheral populations of CD4+ and CD8+ T cells,[36] with the exception of regulatory T cells,[37,38] we cannot exclude the possibility that the absence of CD28 during thymic development has negatively impacted the T cell function. The second salient point is that, as we have already seen in the primary response, the requirement for CD28 costimulation can be impacted by both the duration and strength of TCR stimulation.[57,75] Since LCMV infection overcomes the need for costimulation in the primary, it is reasonable to believe that it may also overcome the need for CD28 costimulation in the secondary memory response. Finally, since both VSV- and LCMV-specific secondary responses have recently been shown to be dependent on the presence of DCs,[86,87] and therefore presumably costimulation, we are led to question findings which indicate that CD8+ T cell memory responses can occur independent of CD28 costimulation.

This brings us to the most current studies, including our own, in which T cell memory, CD4+ or CD8+, is generated in intact mice with acute in vivo viral infections.[85,88] Following the development of intact primary T cell responses, the requirement of the resultant memory population for CD28 costimulation is then studied. This requirement is examined in a variety of ways including the use of CTLA4-Ig, anti-B7 mAb, anti-CD28 mAb, or through transfer of memory cells to CD80/86 double knockout mice (B7 knockouts). Importantly, in each case CD28 knockout mice are avoided, as the lack of CD28 will impair the primary response.

In our studies, virus-specific memory CD8+ T cells were generated through primary infection and 60-day rest in CD28-sufficient mice. This distinction between our studies and previously published work is key. Since the magnitude of the secondary response is dependent upon the magnitude of the primary,[89–91] and the primary response is known to be impaired in CD28–/– mice, it is impossible to determine the role of costimulation in memory CD8+ T cells using CD28–/– mice. Since memory was generated in wild-type C57Bl/6 mice, CD28 costimulation blockade was achieved by treating memory mice with isotype or nondepleting, blocking anti-CD28 mAb[85,92] only during secondary infection. Alternatively, we also deprived memory CD8+ T cells from CD28 costimulation by transferring CD28-sufficient memory to B7 knockout or wild-type control mice, which were then challenged with influenza virus. By delivering CD28-sufficient memory to B7 knockout mice we effectively remove all CD28-mediated costimulation, as well as possible inhibitory CTLA4 signaling from the memory CD8+ T cells. In each strategy employed we measured the re-expansion of virus-specific CD8+ T cells, the functional properties of those cells and the resultant viral loads. This experimental design allows us to overcome some of the confounding factors present in previously published studies and focus exclusively on the role of CD28 costimulation during memory T cell reactivation.

Under the parameters previously described we observed that memory CD8+ T cells require CD28 costimulation to become fully activated during secondary influenza or herpes simplex virus (HSV) infection. Indeed we observed a significant reduction in the re-expansion of memory CD8+ T cells generated against influenza or HSV, when CD28 costimulation is blocked. At the peak of the secondary response of influenza challenged mice, we observed a three-to ninefold reduction in the absolute number of pulmonary virus-specific CD8+ T cells in anti-CD28-treated mice, when compared to untreated or isotype-treated controls. In addition to absolute numbers, a significant reduction in the cytolytic function was observed. The requirement for CD28 costimulation of CD8+ T cell memory was not limited to influenza viral infection alone. To ensure that we were not merely observing a delay in kinetics of expansion in anti-CD28-treated mice, the virus-specific CD8+ T cell population was measured at days 3, 5, 6, 7, 10, and 60 postsecondary challenge. Interestingly, despite the reduction in secondary expansion, the resting secondary memory population on day 60 in anti-CD28-treated mice was equivalent to isotype-treated controls. Future studies will be required to determine whether or not the quality of the secondary memory population is affected in the absence of CD28 costimulation. When HSV-1-specific memory CD8+ T cells were transferred to B7 knockout mice, a significant 4.5-fold reduction in the absolute number of virus-specific CD8+ T cells found in the local draining lymph node was observed, in comparison to controls. Beyond reduced virus-specific CD8+ T expansion and function, we also observed a concurrent increase in viral load. In fact, CD28-blocked influenza, and HSV-infected mice exhibited significantly increased and sustained peak viral loads when compared to their control or isotype-treated counterparts. On day 6 of secondary influenza infection, we observed a tenfold greater viral load in mice which received anti-CD28 treatment compared to isotype-treated controls. The finding of increased viral load reminds us of the critical role that CD8+ T cells play in the elimination of viral infection, and thus the significance of developing efficient CD8+ T cell memory responses.

To examine the mechanism behind this loss of expansion we looked at cellular markers of proliferation and apoptosis. As expected, based on previous findings,[63,64] Bcl-x$_L$ was significantly decreased in our CD28-blocked memory CD8+ T cells in comparison to controls. Surprisingly, a second antiapoptotic molecule, Bcl-2, which rapidly downregulates in activated naïve CD8+ T cells,[93] fails to downregulate in CD28-blocked CD8+ memory T cells when compared to controls. In fact, during a normal activation cycle of a cell, this molecule is downregulated and the cell proceeds into cell cycle.[93] However, if Bcl-2 fails to downregulate, cell cycle is arrested and the cell fails to proliferate.[94–97] Furthermore, cell cycle analysis of virus-specific CD8+ T cells in challenged B7 knockout mice showed that these cells are selectively arrested in the G1/S phase of the cell cycle. Blastogenesis of these cells was not affected, as found earlier.[79] Although these findings do not illustrate a direct interaction, our data do suggest a previously unappreciated relationship between signaling through CD28 and downregulation of Bcl-2.

Similar to in vitro studies of CD8+ memory T cells, the CD4+ T cell memory population, in vitro, has also been suggested to function in a CD28-independent manner.[82,84] However more recently, Ndejembi et al. have provided data which challenges this belief. Here it was shown, using influenza-specific memory CD4+ T cells, that reactivation of memory requires CD28 costimulation. Following in vitro generation of CD4+ TCR transgenic memory or in vivo generation of polyclonal influenza-specific memory CD4+ T cells, the capacity for reactivation was examined. Ex vivo restimulation of memory, in the presence or absence of CTLA4-Ig, led to no discernable differences in the amount of IFN-γ produced by treated or untreated controls. In contrast, IL-2 production by CTLA4-Ig-treated memory CD4+ T cells was significantly decreased in comparison to controls. To examine other functional differences between control and CD28-blocked memory, mice were rechallenged with influenza virus to induce reactivation of memory CD4+ T cells. At the peak of the secondary response, CD4+ memory T cells in recipient mice, which were treated with CTLA4-Ig during influenza rechallenge, exhibited a significant decrease in antigen-specific proliferation in comparison to controls. Taken together, these data strongly indicate that in acute viral infection, CD28 costimulation is required for secondary expansion, and cytokine production by antigen-specific memory CD4+ T cells.

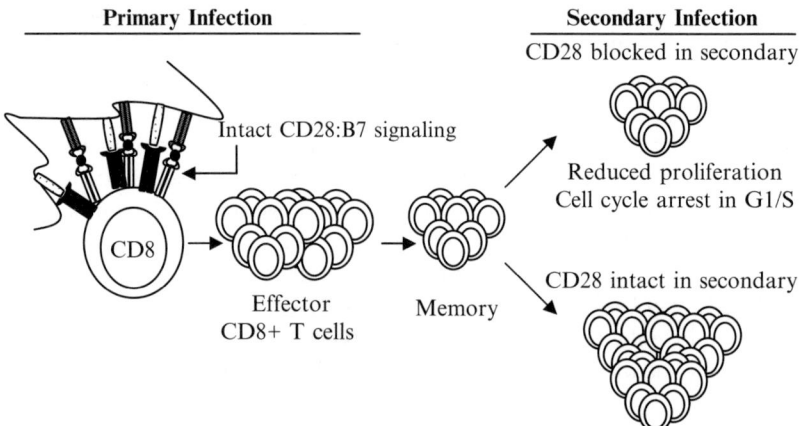

Fig. 3 CD8+ T cell memory responses are CD28 dependent. A study by Borowski et al.[85] indicates that CD28 signaling is required for full expansion and function of antiviral memory CD8+ T cells. This requirement was illustrated using nondepleting, blocking anti-CD28 antibody or transfer to B7 knockout mice during influenza or HSV-1 infection

As we observed with studies examining the functional role of CD28 costimulation in primary antiviral T cell responses, the choice of model system used is critical. The previous assumption that memory T cells do not require CD28 costimulation is based upon in vitro model systems and CD28 knockout mice where the primary response is severely impaired. When memory is generated in CD28-sufficient animals there is a clear requirement for CD28 costimulation to reactivate CD4+ and CD8+ T cell memory[85,88] (Fig. 3).

6 Implications and Future Directions

The implications of the memory T cell requirement for CD28 costimulation are numerous and varied. We can look to several different fields including vaccine development, tumor immunology, autoimmune therapy, or transplantation and easily see how the presence or absence of CD28 costimulation impacts the outcome of a response. One of the immune evasion mechanisms used by viruses, such as measles, Varicella-zoster, and HIV-1, is the suppression of DC maturation and inhibition of CD28 costimulatory ligands, B7-1 and B7-2 upregulation.[98–101] In cases where vaccines may be designed to elicit CD8+ T cell memory against such infections, one might predict a failure to achieve efficient secondary responses due to the absence of costimulation. Based on the previously presented studies one could even predict that the absence of costimulation during reactivation of memory could tolerize the responding memory T cell. Whether this is true, however, remains to be seen. Augmentation of costimulation during reactivation of the vaccine-induced memory T cells may provide the boost needed for successful clearance.

Beyond viral infection and vaccine development we also consider the field of tumor immunology. Like viruses, tumors utilize many different mechanisms to evade CD8+ T cell responses. Some of these mechanisms include the release of anti-inflammatory cytokines, chronic stimulation-induced exhaustion of tumor-associated antigen (TAA)-specific CD8+ T cells, and increased expression of costimulatory molecules with inhibitory function. Specifically, both CTLA4 and the PD-1 ligand B7-H1 are increased in a wide variety a tumors, ranging from ovarian cancer to lung cancer.[102] On the other hand, many studies suggest that DCs may be retained in an immature state, thus not expressing enough CD80 and CD86 to stimulate the CD28 receptor on intratumor T cells.[103,104] While manipulation of costimulation may not address all the tumor evasion mechanisms, it certainly presents an opportunity to enhance TAA-specific CD8+ T cell responses. Indeed blocking of B7-H1 in NOD/SCID mice containing human-explanted tumor tissue resulted in a significant reduction in tumor size.[105,106] Additionally, blockade of CTLA4 has been shown to induce tumor regression. Regression is observed when anti-CTLA4 is given either concurrently with tumorigenic cells or to mice with previously established tumors.[107] Whether the provision of CD28 costimulation, in the form of a soluble ligand or agonist antibody, would augment the CTL response in tumors remains an open question. As with any therapy that functions through blockade of inhibitory molecules, caution must be exercised to prevent the development of autoimmune responses.

Vaccine development and tumor therapeutics represent two areas where the enhancement of CD28 costimulation or blockade of coinhibitory molecules could provide therapeutic benefit to the host. However, in some instances such as autoimmunity and transplantation it is the presence, rather than the absence, of costimulation which may be responsible for inducing undesired immune responses. Therefore, in both autoimmune disease and transplantation, the overarching therapeutic goal is to dampen reactive immune responses against either self or allograft tissue, respectively. Some successes in autoimmune disease have been observed with treatment using CTLA4-Ig which binds B7-1/B7-2 and prevents CD28 costimulation. Studies treating patients with Psoriasis vulgaris[108] and more recently rheumatoid arthritis (RA)[109] show evidence of efficacy. Kremer et al.[109]'s recently published clinical report indicates that RA patients treated with CTLA4-Ig and methotrexate exhibited a significant improvement in clinical score of diseased joints. Interestingly this treatment is even efficacious in those individuals who do not respond to traditional anti-TNFα therapy.[110] The potential use of costimulation inhibition in patients could provide a solution to those who have not benefited from anti-TNFα therapy. However, in light of the recent evidence that the CD28 signaling is essential for the reactivation of memory T cells,[85,88] the use of CTLA4-Ig may also come with undesired consequences such as failure of vaccine or infection-induced memory T cell responses against pathogens.

The field of transplantation has also seen measured success with inhibition of costimulation. In clinical studies by Guinan et al.,[111] ex vivo induction of anergy in allograft bone marrow, prior to transplantation, led to a reduction in the number of patients with graft vs. host disease (GVHD). Other murine studies argue against this finding and indicate that costimulation blockade alone is not sufficient to induce allograft tolerance. Rather a combination of costimulation blockade and NF-κB inhibition is required for tolerance to be induced.[112] Clearly more studies are required to resolve this disparity and others. However, in both transplantation and autoimmunity there is promise that dampening CD28 costimulation can result in a favorable outcome.

Clearly the new understanding of the requirement for CD28 costimulation by memory T cells[85,88] raises important questions about immunity to pathogens and tumors, the effect of costimulation blocking therapies, and the signaling pathway used by memory T cells. Elucidating how CD28 signaling impacts memory T cell responses will further shed light on the role CD28 plays in immune responses.

Acknowledgments This work was supported by NIH grants R01 AI66215, R01 AI46719, and R01 AI62437 from the National Institutes of Health awarded to PDK.

References

1. Bretscher, P. & Cohn, M. A theory of self–nonself discrimination. *Science* **169**, 1042–1049 (1970)
2. Lafferty, K.J. & Cunningham, A.J. A new analysis of allogeneic interactions. *Aust J Exp Biol Med Sci* **53,** 27–42 (1975)
3. Aruffo, A. & Seed, B. Molecular cloning of a CD28 cDNA by a high-efficiency COS cell expression system. *Proc Natl Acad Sci USA* **84**, 8573–8577 (1987)
4. Gmunder, H. & Lesslauer, W. A 45-kDa human T-cell membrane glycoprotein functions in the regulation of cell proliferative responses. *Eur J Biochem* **142**, 153–160 (1984)
5. Ledbetter, J.A. et al. Antibodies to Tp67 and Tp44 augment and sustain proliferative responses of activated T cells. *J Immunol* **135**, 2331–2336 (1985)
6. Moretta, A., Pantaleo, G., Lopez-Botet, M. & Moretta, L. Involvement of T44 molecules in an antigen-independent pathway of T cell activation. Analysis of the correlations to the T cell antigen–receptor complex. *J Exp Med* **162**, 823–838 (1985)
7. Weiss, A., Manger, B. & Imboden, J. Synergy between the T3/antigen receptor complex and Tp44 in the activation of human T cells. *J Immunol* **137**, 819–825 (1986)
8. Martin, P.J. et al. A 44 kilodalton cell surface homodimer regulates interleukin 2 production by activated human T lymphocytes. *J Immunol* **136**, 3282–3287 (1986)
9. June, C.H., Ledbetter, J.A., Lindsten, T. & Thompson, C.B. Evidence for the involvement of three distinct signals in the induction of IL-2 gene expression in human T lymphocytes. *J Immunol* **143**, 153–161 (1989)
10. Ledbetter, J.A. et al. CD28 ligation in T-cell activation: evidence for two signal transduction pathways. *Blood* **75**, 1531–1539 (1990)
11. Linsley, P.S. et al. Coexpression and functional cooperation of CTLA-4 and CD28 on activated T lymphocytes. *J Exp Med* **176**, 1595–1604 (1992)
12. Stamper, C.C. et al. Crystal structure of the B7-1/CTLA-4 complex that inhibits human immune responses. *Nature* **410**, 608–611 (2001)
13. Martin, M., Schneider, H., Azouz, A. & Rudd, C.E. Cytotoxic T lymphocyte antigen 4 and CD28 modulate cell surface raft expression in their regulation of T cell function. *J Exp Med* **194**, 1675–1681 (2001)
14. Walunas, T.L., Bakker, C.Y. & Bluestone, J.A. CTLA-4 ligation blocks CD28-dependent T cell activation. *J Exp Med* **183**, 2541–2550 (1996)

15. Walunas, T.L. et al. CTLA-4 can function as a negative regulator of T cell activation. *Immunity* **1**, 405–413 (1994)

16. Hutloff, A. et al. ICOS is an inducible T-cell co-stimulator structurally and functionally related to CD28. *Nature* **397**, 263–266 (1999)

17. Dong, C. et al. ICOS co-stimulatory receptor is essential for T-cell activation and function. *Nature* **409**, 97–101 (2001)

18. Dong, C., Temann, U.A. & Flavell, R.A. Cutting edge: critical role of inducible costimulator in germinal center reactions. *J Immunol* **166**, 3659–3662 (2001)

19. McAdam, A.J. ICOS is critical for CD40-mediated antibody class switching. *Nature* **409**, 102–105 (2001)

20. Bertram, E.M. et al. Role of ICOS versus CD28 in antiviral immunity. *Eur J Immunol* **32**, 3376–3385 (2002)

21. Keir, M.E., Francisco, L.M. & Sharpe, A.H. PD-1 and its ligands in T-cell immunity. *Curr Opin Immunol* **19**, 309–314 (2007)

22. Barber, D.L. Restoring function in exhausted CD8 T cells during chronic viral infection. *Nature* **439**, 682–687 (2006)

23. Petrovas, C. et al. SIV-specific CD8+ T cells express high levels of PD1 and cytokines but have impaired proliferative capacity in acute and chronic SIVmac251 infection. *Blood* **110**, 928–936 (2007)

24. Trautmann, L. et al. Upregulation of PD-1 expression on HIV-specific CD8+ T cells leads to reversible immune dysfunction. *Nat Med* **12**, 1198–1202 (2006)

25. Petrovas, C. et al. PD-1 is a regulator of virus-specific CD8+ T cell survival in HIV infection. *J Exp Med* **203**, 2281–2292 (2006)

26. Day, C.L. et al. PD-1 expression on HIV-specific T cells is associated with T-cell exhaustion and disease progression. *Nature* **443**, 350–354 (2006)

27. Zeng, C., Wu, T., Zhen, Y., Xia, X.P. & Zhao, Y. BTLA, a new inhibitory B7 family receptor with a TNFR family ligand. *Cell Mol Immunol* **2**, 427–432 (2005)

28. Krieg, C., Boyman, O., Fu, Y.X. & Kaye, J. B and T lymphocyte attenuator regulates CD8+ T cell-intrinsic homeostasis and memory cell generation. *Nat Immunol* **8**, 162–171 (2007)

29. Yang, S.Y., Denning, S.M., Mizuno, S., Dupont, B. & Haynes, B.F. A novel activation pathway for mature thymocytes. Costimulation of CD2 (T,p50) and CD28 (T,p44) induces autocrine interleukin 2/interleukin 2 receptor-mediated cell proliferation. *J Exp Med* **168**, 1457–1468 (1988)

30. Gross, J.A., Callas, E. & Allison, J.P. Identification and distribution of the costimulatory receptor CD28 in the mouse. *J Immunol* **149**, 380–388 (1992)

31. Nelson, A.J., Hosier, S., Brady, W., Linsley, P.S. & Farr, A.G. Medullary thymic epithelium expresses a ligand for CTLA4 in situ and in vitro. *J Immunol* **151**, 2453–2461 (1993)

32. Degermann, S., Surh, C.D., Glimcher, L.H., Sprent, J. & Lo, D. B7 expression on thymic medullary epithelium correlates with epithelium-mediated deletion of V beta 5+ thymocytes. *J Immunol* **152**, 3254–3263 (1994)

33. McKean, D.J. et al. Maturation versus death of developing double-positive thymocytes reflects competing effects on Bcl-2 expression and can be regulated by the intensity of CD28 costimulation. *J Immunol* **166**, 3468–3475 (2001)

34. Cibotti, R., Punt, J.A., Dash, K.S., Sharrow, S.O. & Singer, A. Surface molecules that drive T cell development in vitro in the absence of thymic epithelium and in the absence of lineage-specific signals. *Immunity* **6**, 245–255 (1997)

35. Punt, J.A., Osborne, B.A., Takahama, Y., Sharrow, S.O. & Singer, A. Negative selection of CD4+ CD8+ thymocytes by T cell receptor-induced apoptosis requires a costimulatory signal that can be provided by CD28. *J Exp Med* **179**, 709–713 (1994)

36. Shahinian, A. et al. Differential T cell costimulatory requirements in CD28-deficient mice. *Science* **261**, 609–612 (1993)

37. Tang, Q. et al. Cutting edge: CD28 controls peripheral homeostasis of CD4+ CD25+ regulatory T cells. *J Immunol* **171**, 3348–3352 (2003)

38. Tai, X., Cowan, M., Feigenbaum, L. & Singer, A. CD28 costimulation of developing thymocytes induces Foxp3 expression and regulatory T cell differentiation independently of interleukin 2. *Nat Immunol* **6,** 152 (2005)
39. Walunas, T.L., Sperling, A.I., Khattri, R., Thompson, C.B. & Bluestone, J.A. CD28 expression is not essential for positive and negative selection of thymocytes or peripheral T cell tolerance. *J Immunol* **156,** 1006–1013 (1996)
40. Yamada, H. et al. Monoclonal antibody 9.3 and anti-CD11 antibodies define reciprocal subsets of lymphocytes. *Eur J Immunol* **15,** 1164–1168 (1985)
41. Pellat-Deceunynck, C. et al. Expression of CD28 and CD40 in human myeloma cells: a comparative study with normal plasma cells. *Blood* **84,** 2597–2603 (1994)
42. Kozbor, D., Moretta, A., Messner, H.A., Moretta, L. & Croce, C.M. Tp44 molecules involved in antigen-independent T cell activation are expressed on human plasma cells. *J Immunol* **138,** 4128–4132 (1987)
43. Warrington, K.J., Vallejo, A.N., Weyand, C.M. & Goronzy, J.J. CD28 loss in senescent CD4+ T cells: reversal by interleukin-12 stimulation. *Blood* **101,** 3543–3549 (2003)
44. Effros, R.B. et al. Decline in CD28+ T cells in centenarians and in long-term T cell cultures: a possible cause for both in vivo and in vitro immunosenescence. *Exp Gerontol* **29,** 601–609 (1994)
45. Posnett, D.N., Sinha, R., Kabak, S. & Russo, C. Clonal populations of T cells in normal elderly humans: the T cell equivalent to "benign monoclonal gammapathy". *J Exp Med* **179,** 609–618 (1994)
46. Vallejo, A.N., Weyand, C.M. & Goronzy, J.J. T-cell senescence: a culprit of immune abnormalities in chronic inflammation and persistent infection. *Trends Mol Med* **10,** 119–124 (2004)
47. Inaba, K. et al. The tissue distribution of the B7-2 costimulator in mice: abundant expression on dendritic cells in situ and during maturation in vitro. *J Exp Med* **180,** 1849–1860 (1994)
48. Hathcock, K.S., Laszlo, G., Pucillo, C., Linsley, P. & Hodes, R.J. Comparative analysis of B7-1 and B7-2 costimulatory ligands: expression and function. *J Exp Med* **180,** 631–640 (1994)
49. Frauwirth, K.A. et al. The CD28 signaling pathway regulates glucose metabolism. *Immunity* **16,** 769–777 (2002)
50. Thompson, C.B. et al. CD28 activation pathway regulates the production of multiple T-cell-derived lymphokines/cytokines. *Proc Natl Acad Sci USA* **86,** 1333–1337 (1989)
51. Bjorndahl, J.M., Sung, S.S., Hansen, J.A. & Fu, S.M. Human T cell activation: differential response to anti-CD28 as compared to anti-CD3 monoclonal antibodies. *Eur J Immunol* **19,** 881–887 (1989)
52. Van Lier, R.A., Brouwer, M. & Aarden, L.A. Signals involved in T cell activation. T cell proliferation induced through the synergistic action of anti-CD28 and anti-CD2 monoclonal antibodies. *Eur J Immunol* **18,** 167–172 (1988)
53. Appleman, L.J., Berezovskaya, A., Grass, I. & Boussiotis, V.A. CD28 costimulation mediates T cell expansion via IL-2-independent and IL-2-dependent regulation of cell cycle progression. *J Immunol* **164,** 144–151 (2000)
54. Appleman, L.J., van Puijenbroek, A.A., Shu, K.M., Nadler, L.M. & Boussiotis, V.A. CD28 costimulation mediates down-regulation of p27kip1 and cell cycle progression by activation of the PI3K/PKB signaling pathway in primary human T cells. *J Immunol* **168,** 2729–2736 (2002)
55. Kirchhoff, S., Muller, W.W., Li-Weber, M. & Krammer, P.H. Up-regulation of c-FLIPshort and reduction of activation-induced cell death in CD28-costimulated human T cells. *Eur J Immunol* **30,** 2765–2774 (2000)
56. Sperling, A.I. et al. CD28/B7 interactions deliver a unique signal to naive T cells that regulates cell survival but not early proliferation. *J Immunol* **157,** 3909–3917 (1996)
57. Viola, A. & Lanzavecchia, A. T cell activation determined by T cell receptor number and tunable thresholds. *Science* **273,** 104–106 (1996)
58. Itoh, Y. & Germain, R.N. Single cell analysis reveals regulated hierarchical T cell antigen receptor signaling thresholds and intraclonal heterogeneity for individual cytokine responses of CD4+ T cells. *J Exp Med* **186,** 757–766 (1997)

59. Schwartz, R.H. A cell culture model for T lymphocyte clonal anergy. *Science* **248**, 1349–1356 (1990)

60. Mueller, D.L., Jenkins, M.K. & Schwartz, R.H. Clonal expansion versus functional clonal inactivation: a costimulatory signalling pathway determines the outcome of T cell antigen receptor occupancy. *Annu Rev Immunol* **7**, 445–480 (1989)

61. Harding, F.A., McArthur, J.G., Gross, J.A., Raulet, D.H. & Allison, J.P. CD28-mediated signalling co-stimulates murine T cells and prevents induction of anergy in T-cell clones. *Nature* **356**, 607–609 (1992)

62. Nourse, J. et al. Interleukin-2-mediated elimination of the p27Kip1 cyclin-dependent kinase inhibitor prevented by rapamycin. *Nature* **372**, 570–573 (1994)

63. Boise, L.H. CD28 costimulation can promote T cell survival by enhancing the expression of Bcl-XL. *Immunity* **3**, 87–98 (1995)

64. Wu, L.X. et al. CD28 regulates the translation of Bcl-xL via the phosphatidylinositol 3-kinase/mammalian target of rapamycin pathway. *J Immunol* **174**, 180–194 (2005)

65. Bajenoff, M. et al. Highways, byways and breadcrumbs: directing lymphocyte traffic in the lymph node. *Trends Immunol* **28**, 346–352 (2007)

66. Guermonprez, P., Valladeau, J., Zitvogel, L., Thery, C. & Amigorena, S. Antigen presentation and T cell stimulation by dendritic cells. *Annu Rev Immunol* **20**, 621–667 (2002)

67. Shahinian, A. et al. Differential T cell costimulatory requirements in CD28-deficient mice. *Science* **261**, 609–612 (1993)

68. Suresh, M. et al. Role of CD28-B7 interactions in generation and maintenance of CD8 T cell memory. *J Immunol* **167**, 5565–5573 (2001)

69. McAdam, A.J., Farkash, E.A., Gewurz, B.E. & Sharpe, A.H. B7 costimulation is critical for antibody class switching and CD8(+) cytotoxic T-lymphocyte generation in the host response to vesicular stomatitis virus. *J Virol* **74**, 203–208 (2000)

70. Fuse, S. et al. CD80 and CD86 control antiviral CD8+ T-cell function and immune surveillance of murine gammaherpesvirus 68. *J Virol* **80**, 9159–9170 (2006)

71. Lumsden, J.M., Roberts, J.M., Harris, N.L., Peach, R.J. & Ronchese, F. Differential requirement for CD80 and CD80/CD86-dependent costimulation in the lung immune response to an influenza virus infection. *J Immunol* **164**, 79–85 (2000)

72. Bertram, E.M., Lau, P. & Watts, T.H. Temporal segregation of 4-1BB versus CD28-mediated costimulation: 4-1BB ligand influences T cell numbers late in the primary response and regulates the size of the T cell memory response following influenza infection. *J Immunol* **168**, 3777–3785 (2002)

73. Halstead, E.S., Mueller, Y.M., Altman, J.D. & Katsikis, P.D. In vivo stimulation of CD137 broadens primary antiviral CD8(+) T cell responses. *Nat Immunol* **3**, 536–541 (2002)

74. Liu, Y., Wenger, R.H., Zhao, M. & Nielsen, P.J. Distinct costimulatory molecules are required for the induction of effector and memory cytotoxic T lymphocytes. *J Exp Med* **185**, 251–262 (1997)

75. Kundig, T.M. et al. Duration of TCR stimulation determines costimulatory requirement of T cells. *Immunity* **5**, 41–52 (1996)

76. Sprent, J. & Surh, C.D. T cell memory. *Annu Rev Immunol* **20**, 551–579 (2002)

77. Curtsinger, J.M., Lins, D.C. & Mescher, M.F. CD8+ memory T cells (CD44high, Ly-6C+) are more sensitive than naive cells to (CD44low, Ly-6C−) to TCR/CD8 signaling in response to antigen. *J Immunol* **160**, 3236–3243 (1998)

78. Bachmann, M.F. et al. Developmental regulation of Lck targeting to the CD8 coreceptor controls signaling in naive and memory T cells. *J Exp Med* **189**, 1521–1530 (1999)

79. Kim, S.K., Schluns, K.S. & Lefrancois, L. Induction and visualization of mucosal memory CD8 T cells following systemic virus infection. *J Immunol* **163**, 4125–4132 (1999)

80. Bertram, E.M. et al. A switch in costimulation from CD28 to 4-1BB during primary versus secondary CD8 T cell response to influenza in vivo. *J Immunol* **172**, 981–988 (2004)

81. Flynn, K. & Mullbacher, A. memory alloreactive cytotoxic T cells do not require costimulation for activation in vitro. *Immunol Cell Biol* **74**, 413–420 (1996)

82. Croft, M., Bradley, L.M. & Swain, S.L. Naive versus memory CD4 T cell response to antigen. Memory cells are less dependent on accessory cell costimulation and can respond to many antigen-presenting cell types including resting B cells. *J Immunol* **152**, 2675–2685 (1994)

83. Altman, J.D. et al. Phenotypic analysis of antigen-specific T lymphocytes. *Science* **274**, 94–96 (1996)

84. London, C.A., Lodge, M.P. & Abbas, A.K. Functional responses and costimulator dependence of memory CD4+ T cells. *J Immunol* **164**, 265–272 (2000)

85. Borowski, A.B. et al. Memory CD8+ T cells require CD28 costimulation. *J Immunol* **179**, 6494–6503 (2007)

86. Belz, G.T. et al. Bone marrow-derived cells expand memory CD8+ T cells in response to viral infections of the lung and skin. *Eur J Immunol* **36**, 327–335 (2006)

87. Zammit, D.J., Cauley, L.S., Pham, Q.M. & Lefrancois, L. Dendritic cells maximize the memory CD8 T cell response to infection. *Immunity* **22**, 561–570 (2005)

88. Ndejembi, M.P. et al. Control of memory CD4 T cell recall by the CD28/B7 costimulatory pathway. *J Immunol* **177**, 7698–7706 (2006)

89. Bevan, M.J. & Fink, P.J. The CD8 response on autopilot. *Nat Immunol* **2**, 381–382 (2001)

90. Kaech, S.M. & Ahmed, R. Memory CD8+ T cell differentiation: initial antigen encounter triggers a developmental program in naïve cells. *Nat Immunol* **2**, 415–422 (2001)

91. Murali-Krishna, K. et al. Counting antigen-specific CD8 T cells: a reevaluation of bystander activation during viral infection. *Immunity* **8**, 177–187 (1998)

92. Prilliman, K.R. et al. Cutting edge: a crucial role for B7-CD28 in transmitting T help from APC to CTL. *J Immunol* **169**, 4094–4097 (2002)

93. Grayson, J.M., Zajac, A.J., Altman, J.D. & Ahmed, R. Cutting edge: increased expression of Bcl-2 in antigen-specific memory CD8+ T cells. *J Immunol* **164**, 3950–3954 (2000)

94. Schmitt, C.A. Senescence, apoptosis and therapy – cutting the lifelines of cancer. *Nat Rev Cancer* **3**, 286–295 (2003)

95. Chen, Q.M., Liu, J. & Merrett, J.B. Apoptosis or senescence-like growth arrest: influence of cell-cycle position, p53, p21 and bax in H2O2 response of normal human fibroblasts. *Biochem J* **347**, 543–551 (2000)

96. Sasaki, M., Kumazaki, T., Takano, H., Nishiyama, M. & Mitsui, Y. Senescent cells are resistant to death despite low Bcl-2 level. *Mech Ageing Dev* **122**, 1695–1706 (2001)

97. Kim, R. Unknotting the roles of Bcl-2 and Bcl-xL in cell death. *Biochem Biophys Res Commun* **333**, 336–343 (2005)

98. Servet-Delprat, C. et al. Measles virus induces abnormal differentiation of CD40 ligand-activated human dendritic cells. *J Immunol* **164**, 1753–1760 (2000)

99. Morrow, G., Slobedman, B., Cunningham, A.L. & Abendroth, A. Varicella-zoster virus productively infects mature dendritic cells and alters their immune function. *J Virol* **77**, 4950–4959 (2003)

100. Chaudhry, A. et al. The Nef protein of HIV-1 induces loss of cell surface costimulatory molecules CD80 and CD86 in APCs. *J Immunol* **175**, 4566–4574 (2005)

101. Majumder, B. et al. Human immunodeficiency virus type 1 Vpr impairs dendritic cell maturation and T-cell activation: implications for viral immune escape. *J Virol* **79**, 7990–8003 (2005)

102. Dong, H. et al. Tumor-associated B7-H1 promotes T-cell apoptosis: a potential mechanism of immune evasion. *Nat Med* **8**, 793–800 (2002)

103. Gabrilovich, D.I. et al. Production of vascular endothelial growth factor by human tumors inhibits the functional maturation of dendritic cells. *Nat Med* **2**, 1096–1103 (1996)

104. Gabrilovich, D. et al. Vascular endothelial growth factor inhibits the development of dendritic cells and dramatically affects the differentiation of multiple hematopoietic lineages in vivo. *Blood* **92**, 4150–4166 (1998)

105. Brown, J.A. et al. Blockade of programmed death-1 ligands on dendritic cells enhances T cell activation and cytokine production. *J Immunol* **170**, 1257–1266 (2003)

106. Curiel, T.J. et al. Blockade of B7-H1 improves myeloid dendritic cell-mediated antitumor immunity. *Nat Med* **9**, 562–567 (2003)

107. Leach, D.R., Krummel, M.F. & Allison, J.P. Enhancement of antitumor immunity by CTLA-4 blockade. *Science* **271,** 1734–1736 (1996)
108. Abrams, J.R. et al. CTLA4Ig-mediated blockade of T-cell costimulation in patients with psoriasis vulgaris. *J Clin Invest* **103,** 1243–1252 (1999)
109. Kremer, J.M. et al. Treatment of rheumatoid arthritis by selective inhibition of T-cell activation with fusion protein CTLA4Ig. *N Engl J Med* **349,** 1907–1915 (2003)
110. Genovese, M.C. et al. Abatacept for rheumatoid arthritis refractory to tumor necrosis factor alpha inhibition. *N Engl J Med* **353,** 1114–1123 (2005)
111. Guinan, E.C. et al. Transplantation of anergic histoincompatible bone marrow allografts. *N Engl J Med* **340,** 1704–1714 (1999)
112. Adams, A.B. et al. Heterologous immunity provides a potent barrier to transplantation tolerance. *J Clin Invest* **111,** 1887–1895 (2003)

Peripheral Tolerance Induction by Lymph Node Stroma

Erika D. Reynoso, Je-Wook Lee, and Shannon J. Turley(✉)

1 Introduction

T cell tolerance to self-antigens is of utmost importance to prevent autoimmune attack against self. Failure of T cell tolerance induction inevitably leads to auto-immune disease, as is the case in type-1 diabetes, rheumatoid arthritis, and multiple sclerosis, for example. Fortunately, multiple mechanisms have evolved to enforce self–nonself discrimination among T cells. Such mechanisms act to eliminate self-reactive T cells in the thymus and to eliminate or silence self-reactive T cells, which have escaped thymic selection, in the periphery.

Our current thinking of tolerance induction in the periphery has been shaped by the idea that resting lymph node-resident dendritic cells (DCs) crosspresent tissue-derived self-antigens to $CD8^+$ T cells and incite their elimination. However, recent work by several groups has highlighted a role for nonhematopoietic lymph node stromal cells (LNSCs), with characteristics similar to both medullary thymic epithelial cells (mTECs) and professional antigen-presenting cells (APCs), in the induction of T cell tolerance against self-antigens. In this short communication, we review some of the recently published evidence showing that LNSCs express peripheral tissue-restricted antigens (PTAs), present them on MHC class I molecules, and induce deletion of autoreactive $CD8^+$ T cells. In addition, we also review recent findings that have resulted from the further characterization of the biology of LNSCs. Finally, we provide a discussion on the implications of these findings and some of the questions that warrant further research.

Shannon J. Turley
Department of Pathology, Harvard Medical School and Department of Cancer Immunology and AIDS, Dana-Farber Cancer Institute, Boston, MA 02115, USA

S.P. Schoenberger et al. (eds.) *Crossroads between Innate and Adaptive Immunity II*,
doi: 10.1007/978-0-387-79311-5_10, © Springer Science+Business Media, LLC 2009

2   Mechanisms of T Cell Tolerance to Self

2.1   Induction of Self-Tolerance in the Thymus

Central tolerance mechanisms in the thymus orchestrate the elimination of self-reactive T cells from the developing T cell repertoire. As a result, only 5% of developing T cells are exported as mature cells into the periphery.[48] Negative selection is the mechanism by which self-reactive thymocytes are eliminated in the thymus, and this process is mediated by mTECs and DCs.[19,48] Derbinski and colleagues[16] demonstrated that mTECs express a wide range of PTA genes, *Aire* (autoimmune regulator), functional MHC class II, and the costimulatory molecule B7.1. In addition, by looking at the CD4+ T cell response to serum amyloid protein (SAP) (whose expression is restricted to mTECs) they showed that in wild-type mice the SAP-specific T cell proliferative response was diminished when compared to SAP-deficient mice. This indicated that expression and presentation of SAP by mTECs could lead to tolerance induction of CD4+ T cells. Further evidence for the role of mTECs in tolerance induction came from studies showing that *Aire* expression regulates PTA expression in mTECs.[6] However, because the number of mTECs presenting a given PTA is low, it was proposed that direct presentation by mTECs alone could not account for complete tolerance induction in the thymus.[16,18] Hence, DCs were proposed to also have a role in central tolerance induction. As professional APCs, DCs can present endogenous and exogenous antigens via the classical MHC class I and II pathways, respectively. Furthermore, DCs have the ability to process exogenous antigen and present it on MHC class I molecules to CD8+ T cells, a process termed crosspresentation; this unique ability made it plausible that DCs could have a role in central tolerance in the thymus. In 2004, Gallegos and Bevan showed that DCs can acquire PTAs from mTECs and present the antigens on MHC class I molecules to CD8+ T cells, by crosspresentation, and on MHC class II molecules to CD4+ T cells thereby allowing for clonal deletion of self-reactive thymocytes. In addition, Bonasio and colleagues[11] used mixed bone marrow chimeras, in which a membrane-bound form of OVA was expressed by cardiac myocytes (CMy-mOVA), to demonstrate that tissue-derived DCs could transport CMy-mOVA from the periphery to the thymus and induce deletion of OVA-specific CD4+ thymocytes. Based on their results, they proposed a role for peripheral, migratory DCs in the induction of central tolerance. These results, in addition to the work of others, have led to the proposal of a model in which negative selection of self-reactive thymocytes is mediated by both direct presentation of PTAs by mTECs and by presentation of exogenous antigens acquired by DCs from mTECs or from peripheral tissues (reviewed in ref. [19]).

    In addition to the aforementioned model, the group of Pugliese et al.[20,51] has proposed that DCs in mouse and human thymus (as well as spleen, blood, and peripheral lymph nodes) express transcripts for proinsulin and other type-1 diabetes-associated genes. Based on their data, they proposed that DCs in the thymus and the periphery can express and display, in a transcription-dependent and capture-independent mechanism, multiple self-antigen epitopes to both developing and naïve T cells to regulate both central and peripheral self-tolerance. However, most of their studies are based on the detection of self-antigens on the surface of DCs either by immuno-fluorescence or flow cytometry, and in vivo functional data are yet to be shown.

2.2   DC Crosspresentation in the Induction of Peripheral Tolerance

Despite stringent purging of the T cell repertoire, some self-reactive T cells manage to escape the thymus and enter the periphery, but mechanisms are in place for their elimination and silencing. Control of self-reactive T cells involves two major processes (1) functional suppression or (2) elimination. Functional suppression involves the inhibition of T cell effector function by either (1) regulatory T cells or (2) the induction of anergy or unresponsiveness due to self-antigen presented by APCs under low levels or in the absence of costimulatory molecules. Elimination involves the induction of apoptosis among T cells that respond strongly to self-antigen presented by APCs.[52,58,59] In terms of CD8+ T cell tolerance induction in the periphery, the prevailing paradigm has been that, in the steady state, DCs acquire antigen from parenchymal tissues and traffic to secondary lymphoid tissues where they transfer antigens to lymph node-resident CD8α+ DCs by an as yet unidentified mechanism. Lymph node-resident DCs in turn crosspresent antigen to naive CD8+ T cells leading to an initial burst of proliferation and eventual elimination of self-reactive T cells.[43,52] Kurts et al.[31] provided initial evidence that crosspresentation by DCs had a role in peripheral tolerance induction of CD8+ T cells by showing that DCs could endocytose self-antigen, present it on MHC class I molecules to autoreactive CD8+ T cells, and induce their deletion. This mechanism has been termed crosstolerance. However, the role of DCs in self-antigen presentation has been controversial in terms of their opposing roles in tolerance and immunity, their maturation status, the mechanisms of tolerance induction, and the identity of the crosstolerizing DC. Results from studies looking at the roles of DCs in tolerance vs. immunity have led to the notion that immature or resting DCs mediate tolerance induction whereas mature or licensed DCs drive immunity.[24,40,49,52,58] However, other studies have indicated a possible role for semimature and mature DCs in peripheral T cell tolerance. Semimature DCs are considered to be phenotypically mature (MHC I[high], costimu-lation[high]) but lack the ability to produce proinflammatory cytokines.[35] In addition, semimature DCs have been shown to produce IL-10, induce regulatory T cell pro-duction, and tolerize CD4+ T cells.[2,29,35,39,41] Albert et al.[3] showed that DC maturation is required for crosstolerance of CD8+ T cells and that the immunological outcome of crosspresentation is not determined by the maturation stimulus for the DC. Rather, they proposed that the presence or absence of CD4+ T cell help, which acts in part by delivering a signal to the mature DC via CD40, determine the outcome of crosspresentation: in the presence of CD4+ T cell help the outcome is crosspriming, in its absence the outcome is crosstolerance. Both semimature and mature DCs have been proposed to represent the migratory DCs that constitutively transport antigen from peripheral tissues, such as the skin, to the draining lymph nodes.[3,35] Furthermore, the dynamics of the DC–T cell interactions are different during the induction of tolerance and immunity in peripheral lymph nodes; stable interactions occur during priming whereas brief contacts occur in the induction of CD8+ T cell tolerance.[25]

2.3   Peripheral Self-Tolerance in the Small Intestine

Despite extensive research on peripheral tolerance induction by DCs, most of our current understanding of this process has mainly been based on studies of the pancreas or other tissues which are typically sheltered from environmental exposure.[1,9,30,42] More recently, however, work has been done on the response against self-antigens in the skin using mouse models in which OVA peptide is expressed under the control of different keratin promoters such that OVA is expressed as a self-antigen in the skin.[7,37,38] Surprisingly, these mouse models exhibit severe autoimmune disease in the skin under steady-state conditions, which is driven by CD8+ T cells. Mayerova and colleagues[37] found that CD8+ T cell priming against skin self-antigens is driven by epidermal Langerhans cells. Moreover, when OVA is expressed in all tissues and is presented by other APCs, in addition to Langerhans cells, CD8+ T cell tolerance is the dominant outcome.[37,38] This suggests that there is something unique about tissues that are in contact with the external environment and that additional tolerance mechanisms might be required to prevent autoimmunity.

Until recently, few studies had addressed the mechanisms of T cell tolerance induction in the small intestine, which is in constant contact with the environment.[56,62] The intestine is a particularly interesting locale to study peripheral tolerance since this tissue is constantly bathed in microbial products as well as dietary and environmental antigens, which can potentially activate immune responses. Based on our understanding of DC responsiveness to microbial agents, such environmental factors should make the intestine highly susceptible to autoimmunity; however, due to tightly regulated mechanisms that control immune responses in this vital tissue, autoimmune disorders are not abundant, especially in the small intestine. Currently, DCs are thought to maintain intestinal tolerance by constitutively sampling microbial, dietary, environmental, and intestinal-derived antigens in the steady state and presenting these to naïve T cells in the mesenteric lymph node (MLN) and Peyer's patches.[13,44,45,47,53] Furthermore, DCs in the intestine have been shown to incite production of protective IgA and suppressive cytokines as well as to induce regulatory T cells and T_H2 responses within gut-associated lymphoid tissues (GALT).[36,44,54] As such, intestinal immune responses in the steady state are maintained locally, thereby preventing autoimmunity. Most recently, several groups have described a role for the dietary metabolite retinoic acid in the transforming growth factor-β-mediated induction of FoxP3+ regulatory T cells in the small intestine.[10,14,60] Based on their results, these three groups propose that retinoic acid acts on a specialized subset of CD103[+] DCs within GALT to enhance their ability to induce regulatory T cells as a means of maintaining intestinal self-tolerance.[10,14,60] Furthermore, Denning et al.[15] propose that, in addition to DCs, lamina propria macrophages are also involved in the retinoic acid- and transforming growth factor-β-mediated induction of FoxP3[+] regulatory T cells and include IL-10 as part of this mechanism in the intestine.

3   The Role of Lymph Node Stroma in Tolerance Induction to Intestinal Self

3.1   Lymph Node Stroma Presents Intestinal Self-Antigen in Peripheral Lymph Nodes

Although DCs appear to promote self-tolerance induction in the intestine, an important question still remains. How is tolerance in the small intestine maintained in the constant presence of microbial products and in the presence of inflammation when DCs are so sensitive to these stimuli? To address this question, Vezys et al.[62] generated the iFABP-tOVA transgenic mouse model in which a truncated, cytosolic form of ovalbumin (OVA) is expressed under the control of the promoter for the intestinal fatty acid-binding protein (iFABP) within mature intestinal epithelial cells (IECs). Expression of iFABP is restricted to the IECs of the small intestine and, as such, OVA is expressed as a self-antigen by IECs in the small intestine with highest expression in the terminal ileum. Using this model, they found that naïve OVA-specific (OT-I) TCR transgenic CD8[+] T cells adoptively transferred into iFABP-tOVA mice were tolerized, with most donor T cells being clonally eliminated in lymphoid tissues. From this result, they inferred that deletion of self-reactive CD8[+] T cells was governed by DCs crosspresenting intestinal antigens within GALT.

In light of these results, our laboratory used the iFABP-tOVA model to investigate specifically how self-antigens from the small intestinal mucosa are processed and presented to CD8[+] T cells within secondary lymphoid tissues.[32] Our initial observations indicated that OT-I transgenic CD8[+] T cells, adoptively transferred into iFABP-tOVA mice, not only proliferated efficiently in GALT, as observed by Vezys et al., but also within extraintestinal lymph nodes. Interestingly, proliferation of self-reactive CD8[+] T cells was restricted to peripheral lymph nodes and not observed in the spleen. The possibility that CD8[+] T cells first encountered OVA only in the GALT, began proliferating, and traveled to other peripheral lymph nodes to continue their proliferation was ruled out by showing that the early kinetics of the T cell response were essentially identical in both MLNs and skin lymph nodes (SkLNs) and by showing that blocking lymphocyte egress shortly after antigen exposure had no effect on the T cell response in iFABP-tOVA mice. We also found that presentation of OVA to CD8[+] T cells by DCs is restricted to the GALT and is not observed in extraintestinal lymph nodes; antigen presentation assays demonstrated that in the MLN OVA was selectively presented, but not expressed, by CD8α⁺ DCs. In addition, neither B cells nor macrophages from iFABP-tOVA mice expressed or presented the self-antigen. Most importantly, we found that CD8⁺ T cells directly encounter self-antigen (OVA) within both GALT and extraintestinal lymph nodes on a subset of nonhematopoietic LNSCs comprising 0.05–0.2% of the cell population in all lymph nodes of the body but that is absent in the spleen. Furthermore, LNSCs can process endogenously expressed OVA into functional TCR ligands and stimulate both proliferation and primary activation of naive CD8⁺ T cells in peripheral lymph nodes, but not in spleen.

3.2   *LNSCs Express mRNA Transcripts Encoding PTAs and Can Induce Deletional CD8⁺ T Cell Tolerance*

LNSCs endogenously express mRNA transcripts for PTAs. Initially, this was shown by RT-PCR analysis of lymphoid tissues collected from iFABP-tOVA mice; OVA expression was restricted to nonhematopoietic, CD45⁻ LNSCs and absent from CD45⁺ and CD11c⁺ cells isolated from MLNs and extraintestinal lymph nodes.[32] Expression of PTAs by LNSCs is not restricted only to mRNA for the transgenic OVA antigen since LNSCs from all lymph nodes of the body also express mRNA and protein for multiple intestine-specific PTAs including iFABP and ileal FABP, which mediate intracellular transport and metabolic trafficking of dietary fatty acids, cytokeratin-8, and A33 antigen, a cell surface glycoprotein expressed by mature IECs. In addition, LNSCs express transcripts for other natural PTAs including the eye, thyroid, and central nervous system. Furthermore, more recent data from our laboratory indicate that human lymph nodes express transcripts encoding proteins from the endocrine pancreas and the central nervous system. In addition to the expression of PTAs, LNSCs express transcripts for the autoimmune regulator *Aire*; however, the functional consequence of *Aire* expression by LNSCs is not yet clear. Thus, LNSCs are homologous to mTECs with respect to the expression of PTAs and *Aire*.

The functional significance of PTA expression by LNSCs was demonstrated by showing that LNSCs could process the endogenously expressed self-antigens into functional peptide–MHC class I complexes and present these to CD8⁺ T cells. Ultimately, this led to the deletion of self-reactive CD8⁺ T cells. This was shown by adoptively transferring congenically labeled CD8⁺ T cells into β2m-deficient bone marrow-chimeric iFABP-tOVA mice to restrict antigen presentation to the nonhematopoietic compartment. β2m-deficiency on the hematopoietic compartment was confirmed by showing that chimeric DCs lacked surface MHC class I (H-2Kᵇ) molecules. Analysis of the numbers of transferred T cells after 7 weeks indicated that β2m-deficient bone marrow-chimeric iFABP-tOVA mice had significantly lower numbers of CD8⁺ T cells than control nontransgenic mice. Thus, OVA expression by LNSCs leads to presentation of intestinal antigen to CD8⁺ T cells and results in tolerance induction by deletion of self-reactive T cells. Furthermore, presentation of the gut self-antigen by LNSCs alone is sufficient for tolerance induction and does not require crosspresentation by DCs. This indicates that in the MLNs crosspresentation by DCs may not be necessary for tolerance induction to intestinal self.

More recently, two other groups have described a role for lymph node stroma in the induction of peripheral CD8[+] T cell tolerance. Khazaie and colleagues showed that CD8[+] T cells specific for a transgenic intestinal antigen expressed under the promoter for enteric glial cells underwent primary activation and deletional tolerance in MLNs through interaction with LNSCs expressing and directly presenting the self-antigen in the lymph node cortex (K. Khazaie and H. von Boehmer, personal communication). In addition, they too found that direct presentation of the self-antigen by LNSCs was sufficient to tolerize CD8[+] T cells in vivo and that crosspresentation by DCs was dispensable. Nichols et al.[46] recently published work showing that a population of lymph node-resident, radio-resistant cells express the endogenous melanocyte/melanoma antigen, tyrosinase. Furthermore, using a transgenic mouse model expressing a TCR specific for tyrosinase, they demonstrated that lymph node-resident cells can process endogenous tyrosinase into functional peptide–MHC class I complexes and present antigen to tyrosinase-specific CD8[+] T cells. This resulted in proliferation, primary activation, and ultimate deletion of the self-reactive T cells. In agreement with our work, the Nichols study reported that PTA expression did not occur in the spleen. Interestingly, they found that central tolerance mechanisms were not involved in purging the T cell pool of tyrosinase-specific CD8[+] T cells, despite the fact that tyrosinase was expressed in the thymus. Furthermore, they found that crosspresentation by DCs was not required for peripheral tolerance induction to this endogenous skin antigen. Together, these results suggest that LNSCs play a critical role in establishing and maintaining T cell tolerance to self-antigens.

3.3   Phenotype of LNSCs

Under steady-state conditions LNSCs have a surface phenotype similar to both mTECs in the thymus and professional APCs. LNSCs are selectively tagged with the fucose-specific lectin, *Ulex europaeus* agglutinin-I (UEA-I), which tags mTECs among other APCs in the thymus. Tagging of LNSCs with UEA-I allowed us to determine, using immunofluorescence microscopy, that LNSCs are enriched in the lymph node paracortex and that contact between CD8[+] T cells and UEA-I[+] LNSCs in the cortex is enhanced when the T cell's cognate antigen is directly presented by LNSCs.[32] LNSCs also express gp38 or podoplanin, a mucin-type transmembrane glycoprotein that is expressed by thymic epithelium, lymphatic epithelium, fibroblastic reticular stromal cells in T cell regions of peripheral lymphoid tissues, peritoneal mesothelial cells, osteocytes, glandular myoepithelial cells, ependymal cells, follicular dendritic cells, and ovarian granulosa cells.[12,17,55] LNSCs share other surface markers in common with fibroblastic reticular stromal cells including VCAM-1 and ER-

TR7 (our unpublished observations).[32] Hence, it is possible that LNSCs are in fact fibroblastic reticular stromal cells or a subset of these cells, although this has not been proven definitively.

Like APCs, LNSCs express MHC class II in the form of I-A^b, which raises the possibility that LNSCs may also have the ability to present PTAs to CD4$^+$ T cells. Furthermore, as already discussed, LNSCs can process and present self-antigens on MHC class I molecules to CD8$^+$ T cells, thereby stimulating their proliferation and activation. Due to the similarities between LNSCs and APCs, and to the roles that costimulatory molecules play in tolerance induction by resting DCs, we have recently begun investigating whether LNSCs express costimulatory or coinhibitory molecules found on tolerogenic DCs. Thus far, our data indicate that, under steady-state conditions, LNSCs express low levels of CD80 and CD86 and high levels of PD-L1, but lack surface CD40 and PD-L2 (our unpublished observations). In addition, treatment of mice with the TLR7 agonist, poly I:C, dramatically increases the surface level of PD-L1, but not of other costimulatory molecules or MHC class II, on LNSCs. This is in contrast to DCs, which during proinflammatory conditions, upregulate surface levels of costimulatory molecules, in addition to PD-L1, as well as MHC class II. PD-L1 is a ligand for the programmed cell death-1 receptor (PD-1), which is expressed on T and B cells as well as macrophages.[50,63] PD-L1 is expressed broadly on hematopoietic cells as well as on parenchymal tissues including vascular endothelium, epithelium, muscle, liver, pancreas, islets, placenta, and eye.[28,33] In addition, PD-L1 has been shown to have a predominant role in the inhibition of T cell activation in vivo.[27,28,33,63] Furthermore, Probst et al.[50] showed that resting DCs induce CD8$^+$ T cell tolerance through a synergistic relationship between PD-1 and cytotoxic T lymphocyte-associated antigen-4 (CTLA-4). Using bone marrow-chimeric NOD mice in which PD-ligand expression was restricted to either the hematopoietic or the nonhematopoietic compartment, Keir et al.[28] found that PD-ligand expression on nonhematopoietic cells inhibits self-reactive CD4$^+$ T cells and protects the pancreas from autoimmune attack; this protection was not observed in chimeric mice in which PD-ligand expression was restricted to the hematopoietic compartment. Most importantly, they find that it is specifically PD-L1 expression by parenchymal cells that accounts for the inhibition of CD4$^+$ T cell-mediated destruction of the pancreas and effector cytokine production. More recently, using a murine diabetes model, Keir et al.[27] have shown that PD-L1 expression on parenchymal tissues, but not on hematopoietic cells, can inhibit destruction of the pancreas by in vitro activated self-reactive CD8$^+$ T cells. Thus, our finding that LNSCs express PD-L1 and that surface levels of PD-L1, but not other costimulatory molecules or MHC class II, are increased upon exposure to an inflammatory stimulus indicates that the selective upregulation of PD-L1 on LNSCs might be involved in tolerance induction and maintenance not only during steady-state conditions but also under conditions of inflammation. However, further studies are required to test the functional significance of PD-L1 expression on LNSCs.

4   Implications for the Role of LNSCs in Peripheral Tolerance Induction

4.1   Extrathymic Expression of PTAs in Peripheral Tolerance Induction

Genetic expression of PTAs by nonhematopoietic cells with a role in T cell toler-ance induction had only been demonstrated in mTECs. The finding that extrath-ymic expression of intestinal antigens by nonhematopoietic LNSCs plays a role in displaying gut self to circulating T cells and that LNSCs express multiple PTA transcripts raises the possibility that LNSCs may constitute a highly efficient strat-egy for imposing tolerance in the peripheral T cell pool.[32] Promiscuous PTA expres-sion at two anatomical sites of tolerance induction also raises the question as to whether the same PTAs, or the same level of PTAs, are expressed by both mTECs in the thymus and LNSCs in peripheral lymph nodes. So far, expression of some PTAs has been shown to overlap with mTECs, while others have been shown to be unique to LNSCs. However, the panel of natural PTAs studied in LNSCs is small and further analysis will be necessary to determine the scope of PTA expression by LNSCs and the degree of overlap between the mTECs and LNSCs.

Another important finding is the expression of the autoimmune regulator *Aire* by LNSCs.[32] In the thymus, *Aire* has been shown to act as a transcription factor that regulates the expression of a subset of PTAs by mTECs; loss of *Aire* in the thymus, but not in parenchymal tissues, leads to multiorgan autoimmunity.[6] More recently, *Aire* has also been shown to enhance the ability of mTECs to present PTAs.[5] Furthermore, it was shown that *Aire* does not regulate tolerance induction in the thymus by positive selection of regulatory T cells but through the negative selection of self-reactive effector T cells.[5] In LNSCs, the functional consequences of *Aire* expression have not been determined. Hence, the possibility that *Aire* expression in LNSCs may regulate antigen processing and presentation and/or regulatory T cell induction remains open for LNSCs. Further work will need to be done to determine the role of *Aire* in LNSCs.

4.2   LNSCs vs. DCs in Peripheral Tolerance Induction

An important concept elucidated from the recent identification of LNSCs is that crosspresentation by DCs may be expendable for the induction of T cell tolerance to select self-antigens. In our experimental system, an IEC-associated antigen is crosspresented by CD8α+ DCs in GALT, but not extraintestinal lymph nodes, under steady-state conditions.[32] However, the generation of bone marrow-chimeric mice in which the presentation of peptide–MHC class I complexes was restricted to non-hematopoietic cells allowed us to make the unexpected observation that CD8+

T cells were still tolerized against an IEC-associated protein in the absence of DC-mediated crosspresentation. In a different system, Nichols et al.[46] also determined that CD8[+] T cell tolerance to the skin antigen, tyrosinase, was mediated by a DC-independent mechanism. Both studies came to the conclusion that the induction of peripheral tolerance tracked with a nonhematopoietic cell population present in lymph nodes. The Nichols et al. study also showed that central tolerance mechanisms involving thymic DCs and mTECs were unessential for the deletion of tyrosinase-specific CD8[+] T cells. These new findings beg the question of how extensive the role of LNSCs really is in establishing and maintaining T cell tolerance to PTAs. Future studies will illuminate the relative contributions of crosspresentation by DCs and direct presentation by LNSCs to peripheral CD8[+] T cell tolerance and the molecular mechanisms underlying self-antigen presentation by both APCs.

4.3   LNSCs as Part of the Architecture of the Lymph Node

Another important aspect of the biology and function of LNSCs that warrants further investigation is how these cells relate to the reticular network of the lymph node. This is particularly relevant since, as previously mentioned, LNSCs express gp38, VCAM-1 and ER-TR7, markers associated with fibroblastic reticular stromal cells[12,17,26,55,61] which form the fibroblastic reticular network (FRN) in lymphoid tissues. Furthermore, LNSCs are enriched in the lymph node paracortex. Hence, it is interesting to speculate that LNSCs could be a lineage of fibroblastic reticular stromal cells and/or that they could be associated with the FRN. The FRN is a reticular network associated with the extracellular matrix (ECM) and forms the infrastructure of the lymph node. It has been described as a multifunctional structure that facilitates lymphocyte movement and interaction with FRN-associated APCs, and as a conduit system through which soluble low molecular weight lymph-borne molecules (of less than 70 kDa), cytokines, and chemokines can gain access to the lymph node and high endothelial venules (HEVs).[4,21–23,57] Given the location of LNSCs and their expression of markers associated with the FRN is it possible that LNSCs form part of the FRN conduit? If so, can they process and present soluble antigens that access the lymph node via the FRN? One could envision that LNSCs could be at the interface between the sites of antigen access in the lymph node and that these might process and present lymph-borne antigens to T cells. Hence, it is important to investigate whether LNSCs can endocytose and process lymph-borne antigens, such as soluble self- or dietary antigens that may be coming in through the FRN conduits in an APC-independent manner. Recent work by Bajénoff and colleagues[8] using confocal, electron, and intravital microscopy has shown that stromal cells that form the FRN regulate T cell movement and access the lymph node paracortex, and in addition they show that the FRN defines the limits of T cell movement within the lymph node. More recent work by Link et al.[34] has indicated that a subset of fibroblastic reticular cells, which constitutes the reticular network in the T cell zone of peripheral lymph nodes, can mediate CD4[+] and CD8[+] T cell homeostasis. These "T zone reticular cells" (TRCs) were found to be the main source of IL-7 in mouse peripheral lymph nodes, thereby supporting the survival of naïve T cells. In addition, TRCs were shown to produce the chemokines CCL19 and CCL21, which mediate T cell movement throughout the lymph node and allow for T cells to compete for survival factors, such as IL-7. Our work indicates that CD8+ T cells interact efficiently with LNSCs in the lymph node paracortex. Are the FRN stromal cells described by Baj noff et al. and the TRCs described by Link et al. the same as the LNSCs described by our laboratory? This brings up another important question – do T cells efficiently scan LNSCs as they do DCs in the lymph node? Further work is warranted to investigate the origin and identity of LNSCs and their relationship with the stromal cell network.

5 Summary

In this review we have highlighted the role of LNSCs in the regulation of CD8+ T cell immune responses in peripheral lymph nodes, thereby adding another layer of protection, in addition to the role of resting DCs, against autoimmunity. LNSCs have recently been implicated in the induction of peripheral CD8+ T cell tolerance due to their ability to endogenously express, process, and present PTAs. Furthermore, LNSCs express surface molecules, such as MHC class II and PD-L1, similar to those expressed by mTECs in the thymus and APCs. For future studies it will be important to address some of the new questions that have emerged with respect to the biology and function of LNSCs. Further work will help us to (1) dissect the specific roles that DCs and LNSCs have in the induction and maintenance of tolerance to intestinal antigens, (2) gain a more in-depth understanding of the molecular mechanisms underlying self-tolerance induction by LNSCs and the impact of inflammation on this function, (3) evaluate the relationship of LNSCs to the FRN, and (4) determine if the APC function of LNSCs extends to the acquisition and presentation of exogenous antigens.

Finally, it is important to mention that so far the studies done on LNSCs have focused on their role in CD8+ T cell tolerance. At the moment, we do not know if presentation of PTAs by LNSCs can also induce tolerance of CD4+ T cells. Based on the finding that LNSCs express MHC class II (I-A^b) molecules it is possible that they may present self-antigens to CD4+ T cells and induce tolerance. However, this has yet to be elucidated.

References

1. Adler, A.J., Marsh, D.W., Yochum, G.S., Guzzo, J.L., Nigam, A., Nelson, W.G., and Pardoll, D.M., 1998, CD4+ T cell tolerance to parenchymal self-antigens requires presentation by bone-marrow-derived antigen-presenting cells, *J. Exp. Med.* **187**:1555

2. Akbari, O., DeKruyff, R.H., and Umetsu, D.T., 2001, Pulmonary dendritic cells producing IL-10 mediate tolerance induced by respiratory exposure to antigen, *Nat. Immunol.* **2**:725

3. Albert, M., Jegathesan, M., and Darnell, R.B., 2001, Dendritic cell maturation is required for the cross-tolerization of CD8⁺ T cells, *Nat. Immunol.* **2**:1010

4. Anderson, A.O. and Shaw, S., 2005, Conduit for privileged communications in the lymph node, *Immunity* **22**:3

5. Anderson, M.S., Venanzi, E.S., Chen, Z., Berzins, S.P., Benoist, C., and Mathis, D., 2005, The cellular mechanism of Aire control of T cell tolerance, *Immunity* **23**:227

6. Anderson, M.S., Venanzi, E.S., Klein, L., Chen, Z., Berzins, S.P., Turley, S.J., von Boehmer, H., Bronson, R., Dierich, A., Benoist, C., and Mathis, D., 2002, Projection of an immunological self shadow within the thymus by the Aire protein, *Science* **298**:1395

7. Azukizawa, H., Kosaka, H., Sano, S., Heath, W.R., Takahashi, I., Gao, X.H., Sumikawa, Y., Okabe, M., Yoshikawa, K., and Itami, S., 2003, Induction of T-cell-mediated skin disease specific for antigen transgenically expressed in keratinocytes, *Eur. J. Immunol.* **33**:1879

8. Bajenoff, M., Egen, J.G., Koo, L.Y., Laugier, J.P., Brau, F., Glaichenhaus, N., and Germain, R.N., 2006, Stromal cell networks regulate lymphocyte entry migration and territoriality in lymph nodes, *Immunity* **25**:989

9. Belz, G.T., Behrens, G.M.N., Smith, C.M., Miller, J.F.A.P., Jones, C., Lejon, K., Fathman, C. G., Mueller, S.N., Shortman, K., Carbone, F.R., and Heath, W.R., 2002, The CD8α⁺ dendritic cell subset is responsible for inducing peripheral self-tolerance to tissue-associated antigens, *J. Exp. Med.* **196**:1099

10. Benson, M.J., Pino-Lagos, K., Rosemblatt, M., and Noelle, R.J., 2007, All-trans retinoic acid mediates enhanced T-reg cell growth, differentiation, and gut homing in the face of high levels of co-stimulation, *J. Exp. Med.* **204**:1765

11. Bonasio, R., Scimone, M.L., Schaerli, P., Grabie, N., Lichtman, A.H., and von Andrian, U.H., 2006, Clonal deletion of thymocytes by circulating dendritic cells homing to the thymus, *Nat. Immunol.* **7**:1092

12. Breiteneder-Geleff, S., Soleiman, A., Kowalski, H., Horvat, R., Amann, G., Kriehuber, E., Diem, K., Weninger, W., Tschachler, E., Alitalo, K., and Kerjaschki, D., 1999, Angiosarcomas express mixed endothelial phenotypes of blood and lymphatic capillaries, *Am. J. Pathol.* **154**:385

13. Chieppa, M., Rescigno, M., Huang, A.Y.C., and Germain, R.N., 2006, Dynamic imaging of dendritic cell extension into the small bowel lumen in response to epithelial cell TLR engagement, *J. Exp. Med.* **203**:2841

14. Coombes, J.L., Siddiqui, K.R.R., Arancibia-Carcamo, C., Hall, J., Sun, C.M., Belkaid, Y., and Powrie, F., 2007, A functionally specialized population of mucosal CD103⁺ DCs induces FoxP3⁺ regulatory T cells via a TGF-β- and retinoic acid-dependent mechanism, *J. Exp. Med.* **204**:1757

15. Denning, T.L., Wang, Y., Patel, S.R., Williams, I.R., and Pulendran, B., 2007, Lamina propria macrophages and dendritic cells differentially induce regulatory and interleukin 17-producing T cell responses, *Nat. Immunol.* **8**:1086

16. Derbinski, J., Schulte, A., Kyewski, B., and Klein, L., 2001, Promiscuous gene expression in medullary thymic epithelial cells mirrors the peripheral self, *Nat. Immunol.* **2**:1032

17. Farr, A.G., Berry, M.L., Kim, A., Nelson, A.J., Welch, M.P., and Aruffo, A., 1992, Characterization and cloning of a novel glycoprotein expressed by stromal cells in T-dependent areas of peripheral lymphoid tissues, *J. Exp. Med.* **176**:1477

18. Gallegos, A.M and Bevan, M.J., 2004, Central tolerance to tissue-specific antigens mediated by direct and indirect antigen presentation, *J. Exp. Med.* **200**:1039

19. Gallegos, A.M. and Bevan, M.J., 2006, Central tolerance: good but imperfect, *Immunol. Rev.* **209**:290

20. Garcia, C.A., Prabakar, K.R., Diez, J., Cao, Z.A., Allende, G., Zeller, M., Dogra, R., Mendez, A., Rosenkranz, E., Dahl, U., Ricordi, C., Hanahan, D., and Pugliese, A., 2005, Dendritic cells in human thymus and periphery display a proinsulin epitope in a transcription-dependent, *capture-independent fashion, J. Immunol.* **175**:2111

21. Gretz, J.E., Anderson, A.O., and Shaw, S., 1997, Cords, channels, corridors, and conduits: critical architectural elements facilitating cell interaction in the lymph node corex, *Immunol. Rev.* **156**:11

22. Gretz, J.E., Kaldjian, E.P., Anderson, A.O., and Shat, S., 1996, Sophisticated strategies for information encounter in the lymph node – the reticular network as a conduit of soluble information and a highway for cell traffic, *J. Immunol.* **157**:495

23. Gretz, J.E., Norbury, C.C., Anderson, A.O., Proudfoot, A.E.I., and Shaw, S., 2000, Lymph-borne chemokines and other low molecular weight molecules reach high endothelial venules via specialized conduits while a functional barrier limits access to the lymphocyte microenvironment in lymph node cortex, *J. Exp. Med.* **192**:1425

24. Heath, W.R., Belz, G.T., Behrens, G.M.N., Smith, C.M., Forehan, S.P., Parish, I.A., Davey, G.M., Wilson, N.S., Carbone, F.R., and Villadangos, J.A., 2004, Cross-presentation, dendritic cell subsets, and the generation of immunity to cellular antigens, *Immunol. Rev.* **199**:9

25. Hugues, S., Fetler, L., Bonifaz, L., Helft, J., Amblard, F., and Amigorena, S., 2004, Distinct T cell dynamics in lymph nodes during the induction of tolerance and immunity, *Nat. Immunol.* **5**:1235

26. Kataki, T., Hara, T., Sugai, M., Gonda, H., and Shimizu, A., 2004, Lymph node fibroblastic reticular cells construct the stromal reticulum via contact with lymphocytes. *J. Exp. Med.* **200**:783

27. Keir, M.E., Freeman, G.J., and Sharpe, A.H., 2007, PD-1 regulates self-reactive CD8⁺ T cell responses to antigen in lymph nodes and tissues, J. Immunol. (in press)

28. Keir, M.E., Liang, S.C., Guleria, I., Latchman, Y.E., Qipo, A., Albacker, L.A., Koulmanda, M., Freeman, G.J., Sayegh, M.H., and Sharpe, A.H., 2006, Tissue expression of PD-L1 mediates peripheral T cell tolerance, *J. Exp. Med.* **203**:883

29. Kleindienst, P., Wiethe, C., Lutz, M.B., and Brocker, T., 2005, Simultaneous induction of CD4 T cell tolerance and CD8 T cell immunity by semimature dendritic cells, *J. Immunol.* **174**:3941

30. Kurts, C., Heath, W.R., Carbone, F.R., Allison, J., Miller, J.F.A.P., and Kosaka, H., 1996, Constitutive class I-restricted exogenous presentation of self antigens in vivo, *J. Exp. Med.* **184**:923

31. Kurts, C., Kosaka, H., Carbone, F.R., Miller, J.F.A.P., and Heath, W.R., 1997, Class I-restricted cross-presentation of exogenous self-antigens leads to deletion of autoreactive CD8⁺ T cells, *J. Exp. Med.* **186**:239

32. Lee, J.W., Epardaud, M., Sun, J., Becker, J.E., Cheng, A.C., Yonekura, A., Heath, J.K., and Turley, S.J., 2007, Peripheral antigen display by lymph node stroma promotes T cell tolerance to intestinal self, *Nat. Immunol.* **8**:181

33. Liang, S.C., Latchman, Y.E., Buhlmann, J.E., Tomczak, M.F., Horwitz, B.H., Freeman, G.J., and Sharpe, A.H., 2003, Regulation of PD-1, PD-L1, and PD-L2 expression during normal and autoimmune responses, *Eur. J. Immunol.* **33**:2706

34. Link, A., Vogt, T.K., Favre, S., Britschgi, M.R., Acha-Orbea, H., Hinz, B., Cyster, J.G., and Luther, S.A., 2007, Fibroblastic reticular cells in lymph nodes regulate the homeostasis of naïve T cells, *Nat. Immunol.* (published online)

35. Lutz, M.B. and Schuler, G., 2002, Immature, semi-mature and fully mature dendritic cells: which signals induce tolerance or immunity? *Trends Immunol.* **23**:445

36. Macpherson, A.J., and Uhr, T., 2004, Induction of protective IgA by intestinal dendritic cells carrying commensal bacteria, *Science* **303**:1662

37. Mayerova, D., Parke, E.A., Bursch, L.S., Odumade, O.A., and Hogquist, K.A., 2004, Langerhans cells activate naïve self-antigen-specific CD8 T cells in the steady state, *Immunity* **21**:391

38. McGargill, M.A., Mayerova, D., Stefanski, H.E., Koehn, B., Parke, E.A., Jameson, S.C., Panoskaltsis-Mortari, A., and Hogquist, K.A., 2002, A spontaneous CD8 T cell-dependent autoimmune disease to an antigen expressed under the human keratin 14 promoter, *J. Immunol.* **169**:2141

39. McGirk, P., McCann, C., and Mills, H.G., 2002, Pathogen-specific T regulatory 1 cells induced in the respiratory tract by a bacterial molecule that stimulates interleukin 10 production

by dendritic cells: a novel strategy for evasion of protective T helper type 1 responses by Bordetella pertussis, *J. Exp. Med.* **195**:221

40. Melief, C.J.M., 2003, Regulation of cytotoxic T lymphocyte responses by dendritic cells: peaceful coexistence of cross-priming and direct-priming? *Eur. J. Immunol.* **33**:2645

41. Menges, M., Rössner, S., Voigtländer, C., Schindler, H., Kukutsch, N.A., Bogdan, C., Erb, K., Schuler, G., and Lutz, H., 2002, Repetitive injections of dendritic cells matured with tumor necrosis factor alpha induce antigen-specific protection of mice from autoimmunity, *J. Exp. Med.* **195**:15

42. Morgan, D.J., Kurts, C., Kreuwel, H.T.C., Holst, K.L., Heath, W.R., and Sherman, L.A., 1999, Ontogeny of T cell tolerance to peripherally expressed antigens, *Proc. Natl Acad. Sci. USA* **96**:3854

43. Moser, M., 2003, Dendritic cells in immunity and tolerance-do they display opposite functions? *Immunity* **19**:5

44. Mowat, A.M., 2003, Anatomical basis of tolerance and immunity to intestinal antigens, *Nat. Rev.* **3**:331

45. Nagler-Anderson, C., 2001, Man the barrier! Strategic defenses in the intestinal mucosa, *Nat. Rev.* **1**:59

46. Nichols, L.A., Chen, Y., Colella, T.A., Bennett, C.L., Clausen, B.E., and Engelhard, V.H., 2007, Deletional self-tolerance to a melanocyte/melanoma antigen derived from tyrosinase is mediated by a radio-resistant cell in peripheral and mesenteric lymph nodes, *J. Immunol.* **179**:993

47. Niess, J.H., Brand, S., Gu, X., Landsman, L., Jung, S., McCormick, B.A., Vyas, J.M., Boes, M., Ploegh, H.L., Fox, J.G., Littman, D.R., and Reinecker, H.C., 2005, CX$_3$CR1-mediated dendritic cell access to the intestinal lumen and bacterial clearance, *Science* **307**:254

48. Palmer, E., 2003, Negative selection – clearing out the bad apples from the T-cell repertoire, *Nat. Immunol.* **3**:383

49. Probst, H.C., Lagnel, J., Kollias, G., and van den Broek, M., 2003, Inducible transgenic mice reveal resting dendritic cells as potent inducers of CD8+ T cell tolerance, *Immunity* **18**:713

50. Probst, H.C., McCoy, K., Okazaki, T., Honjo, T., and van den Broek, M., 2005, Resting dendritic cells induce peripheral CD8+ T cell tolerance through PD-1 and CTLA-4, *Nat. Immunol.* **6**:280

51. Pugliese, A., Brown, D., Garza, D., Murchison, D., Zeller, M., Redondo, M., Diez, J., Eisenbarth, G.S., Patel, D.D., and Ricordi, C., 2001, Self-antigen presenting cells expressing diabetes-associated autoantigens exist in both thymus and peripheral lymphoid organs, *J. Clin. Invest.* **107**:555

52. Redmond, W.L. and Sherman, L.S., 2005, Peripheral tolerance of CD8+ lymphocytes, *Immunity* **22**:275

53. Rescigno, M., Urbano, M., Valzasina, B., Francolini, M., Rotta, G., Bonasio, R., Granucci, F., Kraehenbuhl, J.P., and Ricciardi-Castagnoli, P., 2001, Dendritic cells express tight junction proteins and penetrate gut epithelial monolayers to sample bacteria, *Nat. Immunol.* **2**:361

54. Rimoldi, M., Chieppa, M., Salucci, V., Avogadri, F., Sonzogni, A., Sampietro, G.M., Nespoli, A., Viale, G., Allavena, P., and Rescigno, M., 2005, Intestinal immune homeostasis is regulated by the crosstalk between epithelial cells and dendritic cells, *Nat. Immunol.* **6**:507

55. Schacht, V., Dadras, S.S., Johhson, L.A., Jackson, D.G., Hong, Y.K., and Detmar, M., 2005, Up-regulation of the lymphatic marker podoplanin, a mucin-type transmembrane glycoprotein, in human squamous cell carcinomas and germ cell tumors, *Am. J. Pathol.* **166**:913

56. Scheinecker, C., McHugh, R., Shevach, E.M., and Germain, R.N., 2002, Constitutive presentation of a natural tissue autoantigen exclusively by dendritic cells in the draining lymph node, *J. Exp. Med.* **196**:1079

57. Sixt, M., Kanazawa, N., Selg, M., Samson, T., Roos, G., Reinhardt, D.P., Pabst, R., Lutz, M.B., and Sorokin, L., 2005, The conduit system transports soluble antigens from the afferent lymph to resident dendritic cells in the T cell area of the lymph node, *Immunity* **22**:19

58. Steinman, R.M., Hawiger, D., Liu, K., Bonifaz, L., Bonnyay, D., Mahnke, K., Iyoda, T., Ravetech, J., Dhodapkar, M., Inaba, K., and Nussenzweig, M., 2003, Dendritic cell function in vivo during the steady state: a role in peripheral tolerance, *Ann. N.Y. Acad. Sci.* **987**:15

59. Steinman, R.M., Hawiger, D., and Nussenzweig, M., 2003, Tolerogenic dendritic cells, *Annu. Rev. Immunol.* **21**:685

60. Sun, C.M., Hall, J.A., Blank, R.B., Bouladoux, N., Oukka, M., Mora, J.R., and Belkaid, Y., 2007, Small intestine lamina propria dendritic cells promote de novo generation of FoxP3+ T reg cells via retinoic acid, *J. Exp. Med.* **204**:1775

61. van Vliet, E., Melis, M., Foidart, J.M., and van Ewijk, W., 1986, Reticular fibroblasts in peripheral lymphoid organs identified by a monoclonal antibody, *J. Histochem. Cytochem.* **34**:883

62. Vezys, V., Olson, S., and Lefrançois, L., 2000, Expression of intestine-specific antigen reveals novel pathways of CD8+ T cell tolerance induction, *Immunity* **12**:505

63. Yamazaki, T., Akiba, H., Iwai, H., Matsuda, H., Aoki, M., Tanno, Y., Shin, T., Tsuchiya, H., Pardoll, D.M., Okumara, K., Azuma, M., and Yagita, H., 2002, Expression of programmed death 1 ligands by murine T cells and APC, *J. Immunol.* **169**:5538

Index

Printed in the United States of America